T0073614

THE EXQUISITE MACHINE

THE EXQUISITE MACHINE

THE NEW SCIENCE OF THE HEART

SIAN E. HARDING

The MIT Press
Cambridge, Massachusetts
London, England

The MIT Press would like to thank the anonymous peer reviewers who provided comments on drafts of this book. The generous work of academic experts is essential for establishing the authority and quality of our publications. We acknowledge with gratitude the contributions of these otherwise uncredited readers.

This book was set in Bembo Book MT Pro by New Best-set Typesetters Ltd. Printed and bound in the United States of America.

Library of Congress Cataloging-in-Publication Data

Names: Harding, Sian E., author.
Title: The exquisite machine : the new science of the heart /
    Sian E. Harding.
Description: Cambridge, Massachusetts : The MIT Press, [2022] | Includes
    bibliographical references and index.
Identifiers: LCCN 2021052081 | ISBN 9780262047142 (hardcover)
Subjects: LCSH: Heart—Diseases—Treatment. | Heart—Diseases—
    Prevention.
Classification: LCC RC683.8 .H37 2022 | DDC 616.1/2—dc23/
    eng/20211124
LC record available at https://lccn.loc.gov/2021052081

10 9 8 7 6 5 4 3 2 1

This book is dedicated to Ray and Elizabeth Harding

"and this is the wonder
that's keeping the stars apart
i carry your heart
(i carry it in my heart)"
—e.e. cummings

CONTENTS

INTRODUCTION

It was Valentine's Day as I started to write this book, and the world was full of hearts—pink symbols scattered over shops and TVs, emojis strewn over our messages. The heart is our companion in grief or joy, and so has become the inspiration for poets down the centuries. To the poet, the heart is the seat of all emotions: hearts ache, explode with love, they are tender but can harden with rage, they can entwine with others and break with pain. Dominion of the heart is assured. Keats wrote, "I am certain of nothing but the holiness of the Heart's affections"; Emily Dickinson said, "the Heart is the capital of the Mind." We feel a direct line to our emotions through this most familiar of our hidden organs, as we sense it race or pound. Yet the heart is also a marvel of construction unsurpassed by any human creation, honed by 520 million years of evolution. Now that we can observe it more directly, ticking away on the monitors that circle our wrists, we begin to understand its awe-inspiring precision. We see the heartbeats speed up or slow down with our actions and moods, counting down the seconds of our lives. To the engineer, the heart is possibly the most perfect machine ever encountered.

For the past 40 years I have been gazing at this exquisite construction in my laboratory. When I look at the human heart beating in the chest during an operation or lying in a dish when removed for a transplant, it just looks like a glistening

Figure 1.1
A single cardiomyocyte, one-tenth of a millimeter long.

lump of meat. It's hard to associate that solid red muscle with the romantic pink decoration, or the literary description of hearts soaring, bursting, sinking, and breaking. But the new tools of science are revealing a deeper beauty that lies in the perfection of the heart's construction. Once upon a time, I fell in love with a cell. It was a single heart-muscle cell, or cardiomyocyte, and I was one of the first people to work out how to take it undamaged from the dense 3D jigsaw that makes up the heart wall. It lay in the dish under my microscope with a beautiful crystalline structure, a long rectangle with parallel stripes, more like a computer chip than an organic thing. When I added calcium to the dish a long sinuous wave of contraction flowed from one end to the other. When I put electrodes across the dish and turned on the current it beat in perfect time—a tiny heart, the width of a human hair.

The heartbeat may be the first thing we see of our new baby, as even by six weeks after conception this tiny flicker will be visible. During the first ultrasound scan, the doctor will point out the beating heart as a reassuring sign that all is well. We are now used to seeing incredible images in the media of the moving heart, in recordings of open-heart operations or highly detailed scans such as the MRI. The steady upward flicks, then sudden flatline, of the ECG on the heart monitor above a patient's bed is a staple of medical drama. As we age, nearly half of us in the Western world will be treated with some kind of cardiac drug—often blood pressure medications or new agents to control cholesterol and lipids. We search the menus for the "heart-healthy" options and check our heart rate and step count on our mobile devices.

We are right to guard our one-and-only heart with tender care. As a cardiac scientist, I must explain often how dangerous and prevalent heart disease is. And that of course is true: it is the top cause of deaths globally. But I find it more astonishing how many of us don't die sooner: we can survive shocks, injuries, torture, and extreme deprivation without our heart giving up on us. The unrelenting toil of the heart is amazing, and even more so when you realize that half the cardiac muscle cells in your heart survive from the time you're born until you die. A tiny individual cardiomyocyte, barely visible, can pump away diligently for 85 years or more. Contrast this with your skin, which is shed continuously, or your liver, which can regenerate half its cells in a few weeks.

The heart shares with the brain this constancy and stability of construction—and yet both can adapt almost instantly to the challenges of everyday life. The heart can double its output in a matter of seconds or adapt to the enormous demands of supporting a growing fetus. We did not always conceive of the heart as a pump, rather, we have tended to understand its mechanism by analogy with the technology of the day.[1] In ancient times, when smelting furnaces and cooking fires were commonplace, it was held that the purpose of the heart was to warm the blood. It was not until the seventeenth century that the English physician, William Harvey, took the imaginative leap to see the heart as a pump that drives circulation of the blood. This coincided with the increasing use of pumps in industrial processes as their workings became well known. Harvey also understood the electrical pacemaker of the heartbeat in terms of the firing of a musket—a rapid trigger setting in motion a complex series of events. Now, our long history of making machines helps us to understand the incredible precision of the heart's construction with more sophisticated analogies. Our understanding of electrical circuits has combined with new tools to visualize the flow of currents across and within the heart. Our work in robotics has shown us the adaptability and versatility of the cardiac feedback systems, adjusting the blood flow on a second-by-second basis.

The heart is a construction of astounding reliability. Every day it beats 100,000 times and pumps 7,600 liters of blood. Now that about 50 percent of people being born can expect to live to 100 years old, that's more than three billion heartbeats in a lifetime. If you miss more than four minutes together, or 240 of those beats, you will die. Your washing machine spins 18,000 times in a 15-minute cycle. To match your heart's pace, that would be 10 washes per day for 1,000 years. We can send a vehicle to explore Mars, hold the world's knowledge in a phone in our hands, and genetically modify an embryo, but we have not yet been able to build a living heart.

And the heart must be so strong to withstand all the world has to throw at us. I'll show you how we are winning the war against heart attacks, but by surviving with damaged heart muscle we are in growing danger of heart failure. New threats are emerging, like the drugs for cancer treatment that injure the fabric of the heart and leave cancer survivors with a legacy of cardiac devastation. COVID-19 has shown us how violent inflammatory reactions to infection can spread a fire of destruction through the blood vessels of the heart. Pollution, noise, and stress are external threats that we cannot control but have rapid and lasting effects on cardiac function. The poets were right that even the daily emotional onslaught of anguish, grief, and fear take a toll on our poor hearts—I will show how you can literally die of a broken heart.

We need new weapons in the fight against these evolving threats. New technology helps us fight back with imaginative ways of understanding the heart and ingenious techniques to combat cardiac disease. I want to explain the amazing new science of the heart and what vital new information it can offer about your health and your lifespan. Science has given us insights beyond our imagination: taking information from the world too small for the reach of light and from the span of information data banks so large we need artificial intelligence to interpret them. These vast data resources are so sensitive that we can see that the performance of a surgeon on the day after their birthday is measurably worse than on any other day of the year, or that a male cardiologist who has worked with female colleagues will have a markedly better survival rate in his female patients.[2] We can see that a two-hour walk in a busy street creates clear damage to the heart while the same walk in nearby parks improves your health in matter of hours.[3]

In the chapters that follow, I want to share how new scientific developments are opening the mysteries of the heart, so that even scientists like me who have been in the field for our working lives have had to completely rethink our basic framework of ideas month by month. This explosion of new science—ultrafast imaging, gene editing, stem cells, artificial intelligence, and advanced microscopy, below the wavelength of light—has peeled away the layers of complexity to reveal the poetry in motion within the machine. New technology has given us new biological understanding. With our innovations in imaging, we have seen cells talk to each other, exchanging information by intimately embracing or sending packets of information over long distances in the body. From stem cell science, we have discovered that a mother's body can contain stem cells from her children even to very old age, and that these can rush to repair the sites of injury and damage. Using precise physiological measurements, we can detect that the heart is creating emotion as well as responding to it.

These new advances in understanding are not only theoretical, they also have real-world outcomes that can directly help your health. To take one example, we are now able to create new cardiac muscle in a Petri dish. If you give me a few of your cells—say three tablespoons of blood, or a centimeter of skin—I can treat them with four factors and turn back the clock to make them resemble your earliest cells, the first cells that you grew from after the sperm and egg got together and formed every organ in your body. These are the famous *embryonic stem cells*. They can multiply and expand their numbers indefinitely in the Petri dish when we keep them in one brew. If we switch them to other recipes, modeled on the hormones and factors that shaped you in the womb, we can create the cells of heart, liver, bone, and any other organ. We make a dishful of cardiomyocytes, add a pinch of structural cells, and blend in a few other kinds to make blood vessels. Then we suspend them in a gel that molds them all together into a patch of mixed cells. Wait seven days and the new patch we made starts to beat before our very eyes! We call this "engineered heart tissue," strips of beating muscle with all the same genes as you. If your heart has a rhythm problem, then your engineered heart tissue will have one too. These personalized mini hearts in a dish can then be treated with drugs, to find exactly the right one for you!

I hope that this book will show you our increased understanding of an organ that combines reliability with exquisite adaptability beyond that of any machine. It will describe how we are redefining the meaning of what it is to be normal as we unlock the huge wealth of population genetic data. I also explore how our lifestyle interacts critically with our genetic legacy to make us tough enough to survive incredible hardships, or so fragile that a surprise birthday party could be fatal to us. This book will tell you how we are winning the battle with heart disease by harnessing the most cutting edge of the new science—but how the very perfection of the heart is our biggest challenge because it resists our best efforts to invade its precise engineering. This is where we are today, with biology calling on engineering and big data to help solve the puzzles of disease. On the microscale and nanoscale, I talk about new ways to use imaging to track stem cells as they combine with the heart; new models to study the connection between genuine heart muscle and the engineered kind; and new materials to shape the electrical impulses as they threaten the integrity of the tightly packed heart wall. On the massive scale, I show how whole populations of data can help us understand the genome of the individual and the "interactome" of everything that happens to us over our life course. Cardiac scientists must become engineers good enough to match the perfect engineering of the heart, and mathematicians who can understand the data of an entire life.

THREATS TO THE HEART:
A THOUSAND NATURAL SHOCKS

It is very likely that either you or someone you know has been touched by heart disease, because heart disease is the greatest cause of premature death worldwide. Globally, cardiovascular disease (which encompasses coronary heart disease, strokes, and vascular dementia) accounts for 32 percent of deaths, and this varies surprisingly little between developed areas with the United Kingdom at 29 percent, Europe overall at 43 percent, the United States at 32 percent, and Asia at 33 percent.[1] Cardiovascular disease usually runs neck and neck with cancer but outpaces it in people over 75 years old.

For me, the person affected by heart disease was my father-in-law. He had been working in the docks of London, in damp and drafty warehouses. These poor working conditions gave him persistent chest infections and of course everyone smoked in the 1950s. He developed heart failure from the continued strain on his heart due to lung disease and died at the age of 66, just before he was due to retire. Heart failure is not a good way to die. There is a notion that heart disease is the quick and kind death, and for a few it may be true that a drastic heart attack or a sudden massive disruption to heart rhythm is at least quick. But heart *failure* is a gradual deterioration into chronic breathlessness and debilitating fatigue. If a heart attack is like having your chest crushed, then heart failure feels more like drowning or suffocating. I chose to study heart failure after seeing firsthand what this terrible disease did to my father-in-law, and to our family.

Disturbingly, heart failure is growing in most of the world. Currently, about 6.2 million adults in the United States have heart failure and around half of people who develop heart failure die within five years of diagnosis.[2] Heart failure is slow but deadly, with a survival rate worse than most cancers. Like cancer, its progress often builds up as a series of emergency hospital admissions and periods of apparent remission. Gradually we are finding that there are many threats to the heart that push it into failure, some from very recently discovered external harms and others rooted in our deep evolutionary past.

But first the good news. There has been a dramatic fall in the numbers of heart attacks and strokes over the last 50 years, especially in younger age groups (figure 2.1).[3]

This huge change has been driven by the twin engines of prevention and rapid treatment. There have been big shifts in our lifestyle, and the enormous role of public health initiatives cannot be overstated: antismoking campaigns; screening

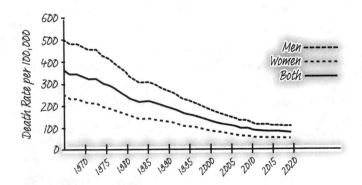

Figure 2.1
Falling death rates from heart disease since 1969.

for and treating high blood pressure, poor blood lipids, and diabetes; encouraging exercise and healthy eating. Results are being seen, and sometimes faster than you might imagine. For example, there was a 2.4 percent reduction in heart attack emergency admissions to the hospital (or 1,200 fewer admissions) in the UK in just 12 months following the ban on smoking in pubs and bars.[4] That is an incredibly rapid effect. Such lifestyle changes have been vital to improving our overall heart health, but also highly effective drugs like statins (to reduce cholesterol) or ACE inhibitors (to lower blood pressure) have made an enormous impact. In clinical trials with more than 70,000 US patients given statins, deaths from myocardial infarction (heart attack) were reduced by 35 percent and from stroke by 30 percent.[5] Statins are now being prescribed to people with cholesterol above a clinical threshold, but who don't yet show signs of heart disease, and this is having a significant protective impact.

We now understand that damage caused by heart blockage mushrooms critically over minutes to hours and, as a result, rapid-treatment protocols or dedicated cardiac centers have been set up in major hospitals. "Time is muscle" as we say. When you call 911 with chest pain the operator knows to prioritize your call and divert help to you as soon as possible. Paramedics in the specially equipped ambulances start the emergency medical procedures straight away. As soon as you arrive at the hospital you are directed to the catheter laboratory to have the blockage removed and a stent put in the offending blood vessel to keep it open. Later, you may have a graft of new blood vessels to replace any blocked ones. These well-designed plans have reduced death rates and improved outcomes for heart attack patients. In the decade up to 2010, survival rates after a heart attack went from 75 percent to 85 percent in the UK with similar changes in the United States.[6]

And yet, despite these heart health improvements, the decrease in death rate as shown in figure 2.1 has begun to level off—and it is not completely clear why. Possibly it is a case of the low-hanging fruit having been harvested: the interventions with the most impact, such as rapid treatments and the availability of drugs including statins and ACE inhibitors, have been put in place. Other behavioral changes such as more exercise, healthier diet, better sleep, will take a while to become embedded in the culture. And there are barriers, both biological and social, that are working to thwart our own best intentions and the efforts of government to make us live healthier lives.

CLASS, STATUS, AND SURVIVAL

Heart disease is clearly linked to social deprivation—one of the explanations for the leveling in the decline of acute cardiovascular disease after 2010 may have been the financial crisis of 2008 and the widespread economic problems that followed. In fact, it is well known that disease of all kinds is linked to poverty. During the COVID-19 pandemic, it became very evident that the outcomes were worse in deprived areas.[7] Even geographically close areas within a large city can have dramatically different death rates across the wealthy/poor divides. Life expectancy in the United States is 10 to 15 years greater for the top 5 percent of earners compared to the bottom 5 percent.[8] At the extreme, homelessness will shorten your life by 30 years.[9]

Some of the damage caused by poverty is linked to behavioral choices, such as smoking, alcohol, or drug abuse and poor diet. But people often blame this disadvantage on lifestyle alone—surely anyone can afford fresh fruit or to take some exercise?—and it is now clear that this is not the whole picture. A constellation of harms that make up a bad environment was defined in a landmark study of Medicare heart failure patients from 440 hospitals in the United States, spanning every kind of geographical and social environment.[10] "Social determinants of health" as they were called, included being black; social isolation (having one or no visits from a family or friend in the past month); no one to care for you during illness; education below high school level; low annual household income (less than $35,000); living in an isolated or small rural area; living in a zip code with more than 25 percent of residences below the federal poverty line; living in an area with little professional health care available. The risk of dying within 90 days of leaving the hospital with heart failure was assessed in nearly 700 patients. Incredibly, the presence of even one of the negative factors was enough to triple the chances of death.

Even more significant, your perception of your place in the social hierarchy can itself effect your health. Studies of British civil servants from the lowest to the highest rank told a very interesting story.[11] These are people with very similar routines and not hugely different salaries. They live a well-ordered life, without extreme excitement, eating similar foods and working in comfortable offices. Yet the highest-ranking civil servants are healthier and live longer. Researchers followed many of these civil servants after retirement and into old age, and the inequalities persisted. Lower grade was strongly associated with slower recovery from poor physical and mental health. While a proportion of the effect on physical health could be explained by differences in healthy behaviors, slower recovery from poor mental health was not accounted for by those differences.[12]

Social class, your place in the social hierarchy, is a statistical determinant of susceptibility to heart (and other) diseases. Historically, this has been the basis for the horrors of eugenics, the idea that the poor are somehow weaker or unfit. But even animals in a highly controlled laboratory environment show the same effect, with high status resulting in a longer life. After months and years at the bottom of the social hierarchy, subordinate mice develop spontaneous atherosclerosis.[13] This is completely absent in their dominant cage mates although both are eating the same food and taking the same exercise. For animals, the effect is much greater in males than females, who do not experience stress from social position to the same degree. The alpha male is a much more widespread phenomenon than the alpha female across the animal kingdom—one study of 76 mammal species with hierarchies found only 7 led by females. Could the same be true for us? Might it contribute to the longer lifespan of women in all cultures? And are changes in women's working patterns the reason that gap is narrowing?

HEALTH INEQUALITY BEGINS BEFORE BIRTH

Threats to the heart are not only from outside—we can have hidden time bombs within ourselves. As we control infectious diseases; make inroads on the extremes of poverty and child mortality; and manage acute heart disease and cancer, our longer and healthier lives reveal a deep-seated genetic lottery. Genetic analysis is now becoming much cheaper and more widespread: so much so that screening of your whole genome will soon be a standard part of medical testing. Although you would think that heart mutations would be fatal, in fact they are surprisingly common. And the ones we know may be the tip of the iceberg.

Scientists study families who have clear signs of an inherited heart condition (enlarged or malformed hearts, erratic heart rhythms) and find the mutation that causes the problem by scanning the genome. A perplexing finding has been that different family members can have the same mutation but very diverse symptoms, with some relatives severely affected and others completely unharmed. Researchers are hunting for the reasons for this variation, and a pattern seems to be emerging that could explain the mystery. A "second hit," or added burden, in either the lifestyle of a particular person or a second genetic variation could be the answer. Although this second hit may have only mild or unnoticeable effects by itself, it can multiply the danger from the mutation. It can make the difference between the way the disease manifests or even, amazingly, whether there is a disease at all!

One of the most striking examples of the effect of "second hits" is for mutations in the protein titin. Titin is the spring that relaxes the cardiomyocyte between beats and is one of the longest proteins in the body when stretched out. Cardiac geneticists studied a population with a form of heart failure unrelated to heart attacks (it's called dilated cardiomyopathy, where the heart wall stretches and thins) and they found that mutations in titin were responsible for the disease in up to a quarter of all patients. It seemed clear that this was a very serious mutation. But when they started to screen large numbers of apparently healthy people, thinking that they would be free of mutations, they had a real shock. One of these studies was done at Imperial College London, where around 2,000 volunteers have had their hearts scanned by MRI and their full genome sequenced. Astonishingly, the researchers found that around 1 percent of the subjects had a titin mutation too![14] Translated to the population of the United States that might mean as many as 2.5 million people with an unsuspected mutation.

With titin as the first clue in the quest, scientists have been seeking the second hit that tips people into disease. This work is very new, but already we are seeing not just one but multiple triggers that reveal the hidden weakness in the heart. Cancer drug treatment is one risk that combines with a titin mutation to make the chances of heart problems higher. Cancer survivors who developed cardiac damage were over 10 times more likely to have a titin variation, and experienced more severe heart disease.[15]

Alcohol is another trigger. A bright spot in the statistics, bucking the trend of relentless self-improvement, has always been the evidence that moderate drinkers are less likely to die from cardiac disease than teetotalers. True, over the years the term "moderate" has gone from "less than your doctor drinks" to a more specific and ever-decreasing number of units, but the basic observation has been hard to shift. However, sadly for us drinkers, the latest research has shown that this rather depends on your genetic makeup. A greater incidence of the titin mutation was found in patients with true alcoholic cardiomyopathy, who had been consuming more than 10 standard drinks a day for years. Most of us are not in that range, so we may think we can relax.[16] But looking at patients with dilated cardiomyopathy of unknown cause, neither having a titin mutation or moderate alcohol consumption above the guidelines alone were significant predictors, but patients with both risk factors had a worryingly large reduction in heart function.[17] The implication here is that these people had been tipped into heart failure through their drinking when others would have been fine.

It's not only chemicals that can cause cardiac stress. Pregnancy obviously is a normal physiological process, but it is not widely appreciated how much it strains the heart. The amount of blood in your body goes up by as much as 50 percent and your heart works harder and beats faster to keep it circulating, to bring blood with oxygen and nutrients to the growing baby. Blood pressure drops as pregnancy hormones make the blood vessels relax and widen. Labor and birth are further physical and emotional stresses, and blood pressure changes during the pushing phase of birth are among the most extreme you will ever experience. It takes several weeks after delivery for the stresses on the heart to return to the levels they were before pregnancy. Some women say they feel shell-shocked after pregnancy and childbirth, and you can see why.

Most women get through this time a bit battered and bruised, but soon bounce back. However, there is a condition known as peripartum cardiomyopathy when mothers unexpectedly develop heart failure late in pregnancy or soon after giving birth. The rate can be as high as 1 in a 100 in certain geographic hot spots, like Nigeria and Haiti, but ranges between 1 in a 1,000 to 1 in 4,000 in Europe and the United States. Mothers are more likely to get peripartum cardiomyopathy if they are having twins, are older or overweight, or have the condition called preeclampsia, where there is a sudden blood pressure rise. Women can recover, but up to 1 in 10 women die and a rising number of heart transplants among women in the United States are to treat peripartum cardiomyopathy. Now we know that hidden mutations are another risk factor for this disease. In one study, women who developed it were six times more likely to have a titin mutation than women in a control group.[18] Once again, the disease was hidden until the second hit of pregnancy and birth occurred.

And titin is just one mutation we know about. Mutations that can cause dilated cardiomyopathy can be found in 1 in 250

people. In fact, for the whole body (not just the heart) it is estimated that each person carries approximately 400 potentially
damaging DNA variants and two mutations known to be associated directly with disease. As many as 1 in 10 people is likely
to develop a genetic disease from these variants. It is sobering
to think that these hidden time bombs in our genes can make
ordinary challenges or choices into unsuspected dangers.

ADDING INSULT TO INJURY

All these causes of damage or dysfunction have the same effect,
to reduce the pumping power of the heart. But it is the body's
own reaction to this loss of cardiac output that deals the final
blow and drives the heart over the cliff into the syndrome
known as heart failure.

Heart failure is not the same as a heart attack. While a heart
attack is among the "insults" that can cause heart failure, the
symptoms are quite different. We should all know the heart
attack symptoms, so I will spell them out here: pain or discom-
fort in your chest that comes on suddenly and doesn't go away,
maybe spreading to your left or right arm, or to your jaw, back
or stomach, as well as feeling sick, sweaty, lightheaded, or
short of breath. This is when you call the emergency services
straight away. Heart failure is not acute but more insidious,
with chronic breathlessness, fatigue and swelling of the ankles
being the key signs.

Heart failure is basically a spiral of destruction, driven by the
body itself through the action of hormones and neurotransmitters, and set in motion by the first damage—be that from heart
attack, drugs, or infection. The body senses the withdrawal
of power from the heart as a lack of blood to the tissues and
immediately reacts to protect itself. However, it misinterprets

the signals because we evolved in a time when humans did not live long enough to get heart attacks and injury instead was the most common threat. In heart failure, the patient's body is reacting as if injury and blood loss were depriving the tissues of blood, and therefore its first goal is to conserve water. The lungs and kidneys work together to produce the hormones angiotensin II and aldosterone, which tighten up the blood vessels and reduce water loss through urine. Adrenaline and noradrenaline stimulate the heart to pump faster and harder, as well as further clamping down blood vessels. The patient's body is flooded with fluid that accumulates around their lungs, making breathing hard, in their gut, disrupting digestion, and in their limbs, causing swelling. No wonder heart failure feels like you are drowning.[19]

A second spiral, only beginning to be understood, is also set in motion. This is low-level immune activation and inflammation that persists after the initial injury. After a heart attack, first there is a dramatic inrush of inflammatory blood cells in reaction to the extreme emergency of devastating cardiac damage. But later, when the damage has been replaced by scarring, a smoldering fire of inflammation persists for months and years. This is the body turning on itself. Usually, our body identifies which proteins are "self" and which are "nonself" and mounts an immune response against the foreign invaders.[20] The sorting function identifying *self* is done before birth in the thymus (a small gland in the throat, also known as sweetbreads in an animal). Anything that is present in the body at this time is checked off as *self* and ignored by the immune system after this point. But some of the proteins inside the cardiac cell are not around before birth; they mature later. When they are released as the heart muscle dies, they are recognized as *nonself* and set off an immune reaction in which antibodies to the heart proteins are made. These attack the heart cells and cause the continuing inflammation and cell death. Big data from large groups of heart attack patients show that the lowest detectable levels of inflammation, once thought to be too small to be dangerous, accurately predict who will die prematurely after heart damage.

All the drugs that we give for heart failure now are to break the vicious cycle of neurohormonal activation, when circulating hormones are raised in the body's attempt to stimulate the failing heart. These drugs include beta-blockers to prevent the action of adrenaline; other blockers also for angiotensin II and aldosterone; diuretics to drive away the excess water; and, of course, drugs to prevent more damage: statins to lower cholesterol; aspirin to prevent blood clots; and antidiabetic agents. These have done well if all you look at is the clinical trials, with each showing around a 10–20 percent improvement in survival over two or three years. New drugs are starting to achieve success in blocking the inflammation and immune reaction, but it's a difficult balance to achieve without leaving the body open to infection.

However, the trials are getting bigger and bigger and the returns smaller and smaller. Each new drug must be tested against the very best combination of the existing drugs. (In fact, it is often said that there is benefit in just being in the placebo group of a clinical trial, to make sure you *do* get the best current treatment.) Every time a new drug is tested the numbers of patients must expand to allow small improvements to be detected, and this makes new cardiovascular trials prohibitively expensive. Big Pharma is backing away from cardiovascular disease as the returns start to dwindle. And in the end, not one of these drugs is a cure. None can reverse the original damage: they simply try to delay the ongoing secondary damage from the body's defenses. None can even stimulate the remaining heart muscle to work harder, or at least not safely.

YOU ARE WHAT YOUR MOTHER ATE

In our evolutionary past, the other big threat apart from injury was starvation. Here again our biology has responded to this potential danger so vigorously that it has paradoxically precipitated one of the greatest perils to the heart and blood vessels that has ever been seen. Diabetes is rising at epidemic rates in most of the world. More than 1 in 10 people in the United States have diabetes: this has tripled since 1980 and is estimated to double again by 2060.[21] The disease itself revolves around blood sugar control. Insulin, which brings glucose (the main blood sugar) into the tissues to harness its energy, is either lacking (as in Type 1 diabetes) or the tissues are insensitive to it (as in Type 2). The result is that blood glucose levels are high. This not only leads to the classic symptom of excess urine production, but it also causes damage of the blood vessels, especially the tiny ones. All tissues have blood vessels, so all are affected by this damage. Blood vessel injury leads to decreased oxygen delivery to the tissue and is responsible for the high level of leg amputations in diabetic patients. But the heart is especially sensitive because of its high oxygen demands—more than two-thirds of deaths in diabetics are from heart disease.

Clues to the reason for the sudden rise in diabetes came from its emergence in different groups of people who had experienced times of hardship. Indigenous populations, like the Australasian Aborigines, and others who have made long migrations, often succumb to diabetes and heart disease in large numbers. The trigger for Type 2 diabetes is certainly an excess of food, obesity, and particularly the Western diet. But the effect of plentiful access to high-calorie meals is much more dramatic in groups of people who have been deprived of food before. More important to note, it is not the starving people themselves who become obese and diabetic, but rather, their children. The child in the womb senses the environment through their mother: he or she prepares for what they will encounter after birth by adjusting their DNA with chemical modifications called epigenetic marks. If their mother is starving during pregnancy, the child will be deprived of nutrition and have a low birth weight. Their genes will be adjusted to put on weight as fast as possible and retain that weight. This is called the "thrifty gene hypothesis." In many different situations, low birth weight is strongly related to diabetes and heart and vascular diseases in later life.

If the child emerges to a world where food is still scarce, then these gene modifications are excellent for their survival. If it is born opposite a shopping mall with 20 different fast-food outlets, then you can see the problem. It's not only biochemical changes to preserve and store calories that are affected by epigenetics, but also behavioral ones like craving high-calorie foods, eating past the point of fullness, and conserving energy by avoiding exercise. The rapid Westernization of large countries like India and China, with food scarcity in their recent historical past, is contributing most to the explosion in diabetes right now. Europe's times of famine are further in the past and so the rise in cases is not quite so rapid. However, epigenetic marks are preserved not only in the children but also the grandchildren of those who have suffered food deprivation, so populations are still vulnerable through multiple generations.

VICTIMS OF OUR OWN SUCCESS

As well as these threats rooted in the depths of evolutionary time, we now have new dangers emerging. Some of these are even from our efforts to treat other diseases: cancer is one example where we have found an unexpected harm to the heart. Curiously, the heart hardly ever gets cancer, so cardiologists and oncologists didn't have any real reason to talk to each other. So it took a little while for it to dawn on the oncologists that their success stories, the people in long-term remission after cancer treatment, were developing heart disease. Of course, both cancer and heart disease become more common as we age, so it was thought that this was just bad luck that someone should get both. It was breast cancer specifically that underlined the fact that this was not just a statistical anomaly but rather, there was a cause-effect relationship between cancer and heart disease. There were two reasons for this. First, the treatment of breast cancer has been remarkably successful. For cancer confined to only the breast, 99 percent of women survive for five years. If the cancer is invasive the survival rates are still a very good 95 percent at five years and 83 percent at 10 years.[22] This means that there are many women now who have been treated with chemotherapy drugs and are effectively cured. Therefore, there was more time for the heart disease to develop and a larger number of people showing these signs of heart disease. Second, this is a female patient group, usually less susceptible to heart disease, and so the unexpected spike in cardiac problems was more remarkable. This alerted the oncologists to look further into the problem. With careful studies and using data from well-controlled clinical trials they discovered that the treatment for breast cancer—as well indeed for any type of cancer—has the potential to damage the heart. To treat one disease was to cause another.

The older generation of cancer drugs are simply cell poisons, blasting everything in their path. They are better at killing rapidly dividing cells, which is why they are so effective on the cancers, but they leave all the body systems with a trail of damage. Your hair is the first casualty of chemotherapy: it is the rapidly dividing cells in the hair follicle that keep your hair growing continuously. The heart muscle itself has very few dividing cells, so that any loss is felt more strongly. And of course your heart, as I shall continue to emphasize, doesn't have a backup. Some of these older (but still very useful) drugs caused heart damage in many patients. At the highest dose used, almost half of the patients treated with one of the mainstay chemotherapy drugs developed heart failure.[23] Radiotherapy, which again largely works by killing the most active cells, is also a danger for heart disease. But we can't throw these therapies away because of the side effects—they are successful in treating the cancer and this must be our priority.

When the newer generations of cancer drugs came along, clinicians hoped that the problem was solved. These drugs attacked the cancer in many ways: some cut off the blood supply to the tumor by interfering with blood vessel development. Others used antibodies, like our natural infection-recognizing molecules, to detect and bind to specific markers on cancer cells. The newest checkpoint inhibitors go even further and use the power of your body's immune system to attack the cancer cells. But the oncologists were deeply disappointed when they discovered that, while these new drugs improved their cancer survival rates enormously, patients were still developing heart disease. Even worse, combining the new with the old drugs could make the effects of the old ones more deadly.[24]

It seems that what is good for curing cancer is bad for the heart and vice versa. Cancer is the most extreme case of cells gone wild, dividing rapidly, and sneaking round the body. The heart is the opposite: the cardiomyocytes are one of the least likely cells to divide or move away from their very firm anchor in the heart wall. Cancer drugs take away the few remaining dividing cells in the heart, reducing its already feeble capacity for repair. Tumors actively attract blood vessels into themselves, to feed their rapidly growing mass. The heart is also a highly vascular organ because of its continuous and enormous energy demands. Some cancer drugs work by blocking blood vessel growth to the tumor, but this also cuts off the blood supply on which the heart is so dependent. It's almost as if, as we age, we are walking a tightrope between the unregulated chaos and expansion of cancer and the degeneration of a moribund heart that is unable to rouse its own cells to divide and save it after injury.

Cancer is not the only success story that has had a downside for the heart. HIV/AIDS has also gone from a death sentence to a chronic condition that can be lived with. However, both the disease itself and, ironically, the drugs to keep it under control can produce heart failure. End-stage AIDS brings with it a severe deterioration of the heart muscle through the many infections caused by the disease, and the large amounts of HIV virus in the blood. But when the new antiretroviral drugs are given, they amplify the conventional cardiovascular risk factors—high blood pressure, poor blood lipid profile, obesity, and diabetes. Combine this with a smoldering undercurrent of infection and inflammation, and the effect is magnified. Myocardial infarction becomes 50 percent more likely; heart failure twice that.[25] Other cardiovascular diseases such as pulmonary hypertension and atrial fibrillation are also seen more often. Countries where HIV/AIDS is endemic, but increasingly under control with drug treatment, are having to gear up their health care systems to handle an explosion of cardiac patient numbers.

For both cancer and HIV/AIDS patients, treating the disease itself is absolutely the top priority. Side effects of the medications must be accepted for now, but scientists and clinicians are putting in a huge effort to head off these problems. All new cancer drugs are tested for cardiotoxicity—the danger of damaging the heart. This is standard for drug development since cardiovascular side effects cause many potential drugs to drop out of the development pipeline before a new product comes to market. Pharmaceutical companies are also seeking new designs and combinations of chemotherapeutic drugs to eliminate cardiotoxicity while keeping the potent anticancer effects. Working from the other side, clinicians are monitoring cancer patients very carefully to catch the first signs of heart disease. Large cancer hospitals are linking with cardiovascular centers to start up clinics where heart and cancer specialists work together to protect the heart while attacking the malignant tumor. Oncologists and cardiologists are getting together and learning one another's language. Looking ahead, cancer patients will be watched closely for their cardiovascular risk factors and may even be treated proactively with heart failure drugs. Guiding the patients between these two serious diseases with conflicting therapies is the new clinical highwire act!

A CONSTELLATION OF COMPLICATIONS

Heart failure has been defined as a syndrome caused by a defect of the heart and recognized by a spectrum of symptoms. You would think from this that it is hard to mistake, but the heart failure symptoms of breathlessness and tiredness can easily be confused with other diseases such as chronic obstructive pulmonary disease (COPD). More than that, heart failure can and often does coexist with COPD and other diseases of age. Multimorbidities, as this is confluence of diseases is called, is the rule rather than the exception for our elderly now. Your aging parent is having trouble walking to the shops—is it just their stiff joints or is it the start of something more serious? Diabetes, fatty liver disease, renal problems, arthritis, dementia are all common and overlapping, with many shared risk factors. Inflammation again is a key underlying part of this condition, driving forward the muscle wasting and frailty that is such a sad feature of aging. Almost half of people aged 65 to 69 years old have two or more chronic conditions and this expands to 75 percent among those age 85 years or older.

Our health systems, which have specialists with deep knowledge of a single organ system (eye, lung, kidney, etc.) actively work against an integrated diagnosis for these patients. Come to the hospital with shortness of breath and it may be a matter of luck whether you are diagnosed with heart failure or COPD—it depends on which consultant is free just then! When we study disease in universities and medical schools it's the same problem of departmental demarcations and scientists from different fields who never meet. We are starting to understand this and fight it with incentives for funding that cross boundaries, buildings designed to encourage the mixing of different specialties, and students who are trained in multiple areas. It's hard work against the institutional barriers and the expanding amounts of information that need to be coordinated, but this must happen to understand disease in a holistic way.

Scientific discoveries are arising in many fields, and they all need to be harnessed to fight the increasing threats to the heart. The basic mechanics of the heart and blood flow are well known, but now we can look further and deeper from the heart muscle to single cells; from single cells to organelles inside cells (like nuclei) with special functions; to nano-domains where individual molecules cluster and interact. Each new advance in the physics of microscopy brings into focus a whole new world of complexity. Chemistry gives us tools to measure the changing distance between molecules as they dance together and apart. To understand the true perfection of the heart we must now visit the realm of objects too small to be sensed by ordinary light.

THE SCIENCE OF THE REALLY SMALL

IT LIVES! IT LIVES!

The human heart is a jigsaw puzzle with five billion pieces, as each of the tiny individual cardiac muscle cells connects with electrical junctions to its neighbors, to conduct the stimulating impulse across the heart. This is a crucial feature, as it lets the heart synchronize perfectly across its whole bulk. It was thrilling when we worked out the exact ingredients to gently tease out these individual cardiomyocytes from a human heart. I remember the intense excitement of the first time I saw these cells, lying in the dish under my microscope. Cardiomyocytes from the main heart wall are separated from their normal pacemaker cells and do not show any movement at first. With bated breath we put electrodes across the dish and turned on the current. The cell twitched and then beat in perfect time—a miniature heart. We couldn't resist a Frankenstein moment—dancing across the lab shouting "It lives! It lives!"

HARNESSING THE POWER OF THE CARDIOMYOCYTE

The most exciting new technologies for cardiac scientists expand our vision on the microscopic level, so that we can appreciate the subtlety and complexity of these individual heart cells. We can see the function of the single cardiomyocyte by how much it shortens (or contracts) with each beat—this directly relates to the force of the heart. The steady pulsation as a heart cell shortens and relaxes is captured and shown on a computer screen—it will stay stable for hours. When we add a stimulant like adrenaline the cardiomyocyte activity increases within seconds, just as it would in your body. This meant we could immediately start to answer many questions, which had been impossible before. For example: Is every cardiomyocyte sensitive to all the heart stimulants and depressants in the same way, or are different areas of the heart specialized? Do cardiomyocytes need to interact with other heart cells to contract most efficiently?

The first question we tackled was: Why can the heart apparently recover from a heart attack, yet start to fail months or years later? We needed to understand what was happening to the individual cells during this time. What was the balance between cell death and poor cell function in the failing heart? We know that billions of cells die during the initial heart attack—is the heart contracting weakly just because it has fewer cardiomyocytes? What about the surviving ones—are they helping by giving maximum performance or adding to the problem by underperforming? To answer these questions, we conducted an experiment that studied cardiomyocytes from failing hearts that were removed during transplants. It was an all-consuming time for our team. Many nights were spent with our phones on 24/7 alert for a call, and then journeying to the hospital in the early hours of the morning. Transporting the precious failing heart back to the laboratory, we worked around the clock to extract the cardiomyocytes and learn their secrets. Gradually the answer to our first question became clear. Some of the cardiomyocytes in the tissue were dead, but live ones remained. The strain on these remaining live cardiomyocytes had changed their cellular organization and caused them to become sluggish and feeble. They contracted slowly and poorly, their relaxation was long and often incomplete, and they frequently fired off irregular beats. Importantly, we observed that this always happened in the same way, independent of how the first injury to the heart had happened—be it a heart attack, valve disease, or genetic problem. It did not depend on the type of damage. This new insight showed us that we can revive a failing heart not only by adding new cells, but also with drugs and gene therapy that target the molecular changes in the remaining malfunctioning cardiomyocytes.

The heartbeat needs to squeeze out blood strongly enough to circulate around the whole body, so the heart muscle must generate strong and rapid force. However, it's also very important for the heart to relax quickly in between beats; otherwise there is not enough time for it to fill with blood before the next one. Our malfunctioning cardiomyocytes, which control the heart force, were dangerous because of their faults in relaxation as well as contraction. The heart must expel its blood rapidly and then relax to fill with more blood—and this happens quicker as your heart rate rises. The faster it can do this, the more efficient the beat. It is easy to see that problems with the force of contraction of the heart will reduce its efficiency in ejecting blood. However, we now realize that many people have problems with their hearts mainly because the *relaxation* is slow, not because the contraction is poor. Heart failure due to inefficient relaxation is a new epidemic and it has links to obesity, diabetes, hypertension, and other chronic conditions such as lung and kidney disease.[1]

Working with individual cardiomyocytes also opens a huge variety of new methods to probe the biology of contraction.

Impaling the cardiomyocyte with a sharp glass electrode tip filled with conducting solution can measure the electrical current change, called the action potential, that happens with each beat. Channels and pumps in the cell membrane that control the movement of sodium, calcium, and potassium salts turn on and off to shape the electrical impulse. Smoothing the tip of the electrode lets us clamp it on the surface of the cell to capture and isolate a minute patch of membrane. Getting the tip into position is a tricky and delicate maneuver using hand-held controllers—video gaming is excellent training for the scientists who do these experiments!

The first thing we see seems like electrical noise, a jumpy random squiggle on our recording equipment. But hidden within the scratchy up-and-down lines is a rich harvest of information. Now we must ramp up the sensitivity of the system to pick out the impossibly tiny currents made by an ion channel, a portal made by a single protein molecule that lets one electrically charged sodium, calcium, or potassium ion at a time into the cell. The channel molecule constantly changes its shape, flickering between open and closed states thousands of times a second. As the ions cross during the brief openings, they create microcurrents. These intricate methods are important to help us understand the electrical activity of the heart, and to develop drugs to keep it stable and healthy. Drugs like calcium channel blockers, for example, act at a molecular level to reduce the time that the calcium ion channel stays open, and thus damp down any dangerous racing heart rhythms.

SETTING THE RHYTHM

Not all cardiomyocytes are created equal. If we take cardiomyocytes from the main muscle of the heart (the atria or

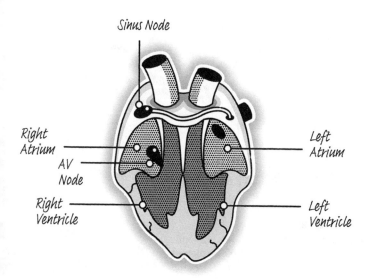

Figure 3.1
The anatomy of the heart. The electrical impulse runs from the sinus pacemaker node to the atria and atrio-ventricular (AV) node and from there to stimulate contraction of the ventricles.

ventricles; see figure 3.1) we see cells with a long stable contraction and then a fast relaxation. They lie still in our Petri dish and only twitch if we pass an electric current across the salt solution in which they are kept. But if we take cells from the small, specialized pacemaker regions of the heart (sinus node, AV node; see figure 3.1) they twitch or beat by themselves. The mechanism for this is a fluctuating electrical current across the cardiomyocyte outer membrane. In the atrial and ventricular cardiomyocytes, the pumps and channels hold a steady voltage difference between inside and outside the cell when they are not actually beating. In the pacemaker cells, one of the channels has an activity that gradually changes with time and makes the pacemaker cells beat spontaneously and

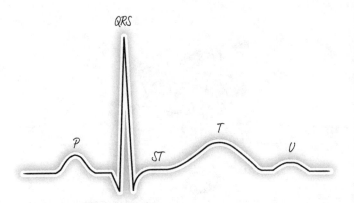

Figure 3.2
The cardiac electrogram (ECG): sodium, potassium, and calcium ions
moving into and out of the cardiomyocytes create tiny electrical
currents that add up to the total current, which starts the heartbeat.
The currents are sensed on the surface of the body by ECG electrodes.
The letters show the progression of the electrical impulse across and
down the different chambers of the heart.

rhythmically. These tiny changes in currents within individual cells add up to the familiar electrocardiogram (or ECG/
EKG; figure 3.2), which is the main measurement of whole-
heart electrical activity.

THE ORCHESTRA PLAYS!

Now that we have taken apart the heart jigsaw puzzle, we
must put it together again to see how the individual cardiomyocytes each play their part in the orchestrated contraction and
relaxation of the heart. In the culture dish, the size and shape
of the electrical signal varies greatly from cell to cell. Adding
drugs that can disturb the electrical signal causes the cell to fire
off irregular beats, or even to writhe uncontrollably. If this

happened in your heart, it could result in dangerous or even fatal disturbances of rhythm, called arrhythmias. But when they are in the muscle of the heart, connections between the cardiomyocytes on all sides stabilize the system: the isolated cells become synchronized with each other and beat more regularly. If we measure cardiomyocytes within an intact strip of heart muscle, where they are fully integrated with their neighbors, the cells behave much more consistently and are more difficult to disturb. Cardiomyocytes in an intact muscle strip can tolerate much more of the arrhythmia-inducing drugs than the individual cells in a dish. In effect, the heart functions as one giant cardiomyocyte, and this gives it a robust stability against electrical disturbance.

The start of the music is signaled by the pacemaker cardiomyocytes, as they spontaneously fire off their regular electrical impulses. The impulse starts at the main pacemaker (the sinus node) on one of the upper chambers of the heart, the right atrium (figure 3.2). The normal rhythm of the heartbeat is called sinus rhythm. (There is another, slower, backup pacemaker above the lower chambers—the ventricles—called the atrio-ventricular or AV node: this can take over if the main pacemaker fails). In the ECG the P wave is when the impulse is going through the atria, which contract first. Then the large spike is the QRS complex that shows the stimulation of the right and left ventricles. The ST segment is the gap between the contraction impulse and the T and U waves, which represent restoration of resting membrane electrical levels by repolarization (the opposite of depolarization). If you are having a heart attack, it is often a rise in the ST segment that gives this away to the clinician.

This superb piece of engineering allows precise coordination of the flow of blood around the lungs and body. Blood comes from the body with low oxygen levels, as it has supplied all the needs of the body organs. It enters through the right atrium, which contracts first and gives it a boost into the right ventricle. From there the blood leaves the heart for a circuit through the lungs, where it is replenished with oxygen. Blood returning from the lungs goes to the left atrium. It is expelled from there it into the main chamber of the heart, the thick and muscular left ventricle. This chamber has the force to drive the blood with each heartbeat into and around the body, against all the resistance of the vessels branching into each organ. The right and left ventricular circuits take place in synchrony with each other in every heartbeat. Blood volumes going through the two circuits must be precisely matched, as even tiny differences would add up very quickly to unbalance the system. As the blood leaves them, each of the chambers of the heart relaxes, ready to receive the new load. The heartbeat is over, that movement of the symphony is complete.

THE THEATRE OF BIOLOGY

Heart science is excitingly visual. I must remember not to hold my breath when studying a beating cardiomyocyte—I've just added a new stimulant drug I'm testing: Will it work? Down the microscope, I can see within seconds the beats getting larger and faster before my eyes. The computer output records the beat so we can examine the contraction and relaxation in detail later. It's much more immediately satisfying than a week of protein extraction or a month of mathematical analysis.

Not content with the beating, we can even make the cardiomyocyte light up in brilliant color. Brightly hued fluorescent sensors have given us entry into the inner world of the cell. In the simplest method, we soak the cells with a compound that increases green fluorescence when a particular molecule—often calcium—rises. This compound is lipophilic, or fat-loving, which allows it to cross the fatty outer cardiomyocyte cell membrane and enter the cell. When the cell beats, the calcium rises and there is a flash of green light. The strength and speed of this signal is very valuable information because calcium is the main driver for cardiomyocyte contraction.

One of the earliest important discoveries showing this was made by the clinician and physiologist, Sydney Ringer, in the 1880s. He was working on hearts removed from frogs and his experiments were going well. Then he was horrified to discover that his technician was using London tap water rather than pure distilled water to keep the hearts. However, when this was corrected, all his experiments failed because the hearts stopped beating. He tracked the vital ingredient down to the calcium or lime content of London water, which to this day is notoriously high in minerals. By chance, it has the same calcium content as blood! This serendipitous mistake by his technician had produced a culture medium that was ideal for his experiments.

Another set of fluorescent compounds senses the electrical potential in the membrane. This can show us how the electrical impulse changes with time in pacemaker cells or becomes disrupted during arrhythmia. Here we want a fat-loving compound to stay in this fatty layer and glow yellow or red as the impulse streaks across. If we are clever with our choice of colors, we can measure both calcium and electrical activity at once in a single cell. A similar technique allows us to see how proteins move and cluster within the cells, and how this changes their function. By labeling each protein with different-colored fluorescent molecules we can keep track of them. We can even understand how their function changes by watching their fluorescence color alter as they approach each other. The images produced are not only valuable for their information, but often very lovely. There are competitions for the most beautiful scientific images run every year by the charities that fund the research, and the cardiac scientists have a big advantage here.

In this way, color allows us to understand the role calcium and proteins play in the mechanism of cardiomyocytes. However, the limitations imposed by the wavelength of light prevents us from observing the very smallest of cellular components, so we have developed technologies to dive even further into the microscopic realm.

SEEING WITHOUT LIGHT

We used to think that the cell was a big bag of liquid surrounded by a membrane, with all the different internal structures (organelles like the nucleus or mitochondria) randomly floating and jostling in a soup of the protein molecules that control their function. This is just one example of how scientists have embarrassingly underestimated the sophistication of biological organization. Actually, the protein molecules are kept in place by both physical and chemical harnesses. They can be bound to an organelle within the cell, like the nucleus, or to a specialized area like the outer cell membrane compartment, only moving after they get a message to act. Or they can be within a group of other molecules that interact to increase or decrease their function. Signaling proteins that produce a messenger molecule are often clustered with others that amplify or dampen that signal and so continuously adjust the final signal strength.

To dive down into this microstructure, we first need to tether our sensors in specific positions. Protein address labels help to do this: small sequences that will bind our sensors to the right cluster and monitor the local environment there. We also had to up our game with the microscopy. And because these molecular couplings are changing and reforming many times a second, we had to do this in live cells.

Our first piece of equipment was quite literally built from parts salvaged from a dustbin. A brilliant young Russian researcher, Yuri Korchev, had the idea to scan the cell with a probe made from one of our electrodes, but without touching the cell surface. Using a discarded microscope he had salvaged from the laboratory junk pile, he built a moving platform for the cell and linked this to an electrode fixed a microscopic distance away from the cell surface. The platform scans backward and forward, moving up and down to keep the gap between probe and cell constant. Scanning the whole cell this way gave us a complete contour map of the surface (figure 3.3).

There were two really clever aspects to this invention, now called scanning ion conductance microscopy (SICM).[2] First, it didn't need light—it was just sensing distance between the cell surface and the probe by electrical charge. This meant it could detect structures below the wavelength of light and show us their shape. Scanning the cardiomyocytes suddenly revealed that what looked like stripes with a light microscope were an alpine landscape of hills and valleys, with deeply penetrating tunnels. We realized there was a whole new world down there to explore. Second, the probe could then find its way back to a particular place or structure and record the local effect. We could steer it toward the fixed sensors in the specific features on the membrane such as the tunnel entrances. Dipping the probe into the mouth of the tunnel let us squirt drugs into the

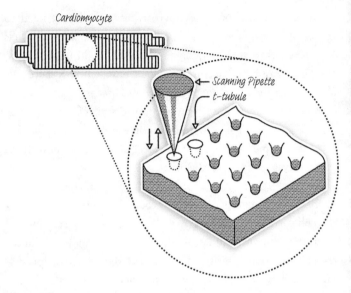

Figure 3.3
In scanning ion conductance microscopy, the scanning pipette moves
back and forth across a small area of the cell to create a contour map of
the surface.

space there. We could then switch back to the light detection
and see the fluorescence changes in our tethered sensors. We
saw that our drugs light up the sensors close to the tunnel but
not at the other end of the cell. It showed us that the signaling
proteins for those drugs were held tightly in the tunnel area
and had a very local range of action, less than 1 percent of the
cardiomyocyte.

When we compared normal human cardiomyocytes with
those from a failing heart, we could immediately see that disease
had devastated the normal landscape. Gone were the undula-
tions and hollows. Patches of flattened membrane spread across
the surface of the cardiomyocyte with only the occasional

tunnel entrance remaining. Now, spraying in the activating drug lit up the whole cell from end to end. The signaling molecules had been released and were wandering throughout the cardiomyocyte. By being in the wrong place and hitting mistaken targets they were disrupting the smooth working of the cell and, by extension, the heart.

I'm pleased to say this worked out well for the dustbin-raiding researcher—Yuri wrote to a senior professor in my department at Imperial College London, who liked the idea so much he fought for Yuri to have a job and he is now a professor here too.

THE LONELY CARDIOMYOCYTE

A sadder job we have is to understand how our cells become injured and expire, but this is vital for the understanding of disease. There are many ways for a cell to die, controlled processes that smoothly absorb the cells that are not needed (which is why you do not have webbed feet) or catastrophic sudden death where a cell explosively loses its contents through a ruptured outer membrane. In a heart attack it is the catastrophic kind—up to two billion cardiomyocytes in the heart wall burst and shrivel with the sudden lack of oxygen. But cells can also die for lack of company—there is a special word for this, *anoikis*, which means loneliness or homelessness. An isolated cardiomyocyte in a dish very soon starts to lose its very precise structure. Within days it regresses to its immature form, rounder and smoother, with the stripes now disorganized. Many other cell types will grow happily outside the body, if they are in a dish bathed in a warm culture medium full of glucose and nutrients. They will divide and proliferate, increasing in number until they fill up the floor of the dish.

Cardiomyocytes never do this—they rarely divide but just gradually atrophy or waste away.

This is true in the whole living heart too, which is why a heart attack is so disastrous. After this dramatic and whole-sale cell death, the remaining cardiomyocytes may divide and increase in number slightly, but nowhere nearly enough to recreate the billion cells needed to repair the damage. (On the positive side, we think this is why it is very rare to get cancer of the heart, since cancer is a disease where proliferation runs out of control.) Instead, there is a takeover from cells like the fibroblasts, which make up the connective tissue holding the heart together. Fibroblasts are naturally very proliferative—the weeds of the cell world—and surge in growth to fill the gaps. They create the scar tissue that is essential to hold the wall of the heart together as the cardiomyocytes die. However, fibroblasts are not contractile and in the long term this scar will stiffen the heart and disrupt the flow of electrical impulses through the muscle. It will contribute to the slow relaxation and filling, which is so harmful to the heart function.

Why do cardiomyocytes hate to be alone so much? We know they miss the continuous mechanical stimulation of the stretch and relaxation cycles that they experience in a living heart. New heart assist devices (partial artificial hearts), which move the blood out of the heart without needing it to beat, decrease the mechanical load on the cardiomyocytes. Very quickly (days to weeks) the heart atrophies and the cardiomyo-cytes become thinner and weaker.[3] In contrast, when we have made new immature cardiomyocytes from stem cells, we push them to become like adult cells by stimulating them to con-tract against a load. If we take ultrathin slices of human heart, we can preserve the function of the cardiomyocytes within by putting them in a bioreactor that continually stretches and stimulates them.

THE MARVELOUS MACHINE

What the cardiomyocyte has taught us, over and above any insights about molecular mechanisms, is that the heart is more than the sum of its individual muscle cells. The cardiomyocytes within the heart are being protected and reinforced by their incredibly close connections with each other. The mechanical stimulus distributed across the cells in the heart wall is continually exercising them to preserve their optimum function. Electrical buffering, a direct result of the multitude of connections on each cardiomyocyte, damps down irregular beats. It is not until many individual cells are arrhythmic that the heart muscle begins to feel the effects and become arrhythmic itself. The heart is effectively behaving as a single cell with an unwavering purpose to sustain the output of blood to the body.

These adaptations, which the heart has made over millions of years of evolution, also underlie its vulnerability and its resistance to repair. Cardiomyocytes do not divide and proliferate to mend the damaged heart because they must not. When a cardiomyocyte divides, it must break down its structure to split into two, and then build it up again before it can rejoin its place in the jigsaw puzzle. If a significant number of cardiomyocytes are in the process of doing this, then they are not contributing to the team effort. Not only that, but they also are a weak spot for the electrical impulse, disturbing its smooth passage across the cardiac wall. So, regeneration represents a danger period for the heart, and we must take great care in our efforts to stimulate it. When we build machines or living grafts to try to repair the heart, we must make something just as perfect, and which can immediately integrate into the existing structure perfectly as well! You can start to understand the huge challenge that we face in attempting to compete with evolution.

We have seen the science of the really small and how this can give us new insights into the heart at the molecular level. Now we turn to the science of the large. One new weapon we have in our armory is big data: public health information on lifestyle, pollution, and the environment; large clinical datasets on whole countries or people tracked from birth to death; and biobanks full of tissue and blood samples. We will explore how data science, with the help of remote monitoring and artificial intelligence, is helping us toward the integrated and inclusive analysis of our lifetime's genetic and environmental influences.

BIG DATA—MANY HEARTS THAT BEAT AS ONE?

Science can be a bit of a bruising business—disagreements about a hypothesis or experimental result can become very heated. Maybe researchers from one university can't repeat the result of another. Or drug companies are complaining that they don't get the same results as university researchers. Murmurs of incompetence (or worse) come from both sides and rational debate deteriorates into name calling. There can be plenty of genuine reasons for a lack of reproducibility in experiments: animal responses vary with the season; cell lines can alter in culture. Sometimes we just don't know why one scientist will have "green fingers"—their experiments just seem to work—and another doesn't. After all, everyone's pastry is slightly different, and that's just flour, butter, and water. I suspect that every scientist has had the experience of not even being able to repeat their *own* results, for reasons they never discover! But critically, we are beginning to understand that one key reason for the disputes and conflicts is that studies from individual labs are just too small. They do not have the statistical "power," or sheer force of numbers, to understand the effects of very small but vitally important biological differences. If you toss a coin four times, you have a reasonable chance of getting heads every time. If you toss it 400,000 times, the chance of heads will be almost exactly 50 percent. This is a physical law—the law of large numbers.

This power effect is glaringly obvious in the nature-vs.-nurture genetic studies, which try to tease out which of our features are laid down in our genes and which are formed later by our environment. The excellent book *Blueprint* by Robert Plomin describes how scientists knew that certain genetic traits like intelligence, mental problems, and weight were strongly inherited.[1] Identical twins who had been adopted by different families almost completely resembled their biological parents and not their adoptive ones. But when they tried to find the genes involved, individual laboratories were coming up with very different answers. It took much larger groupings, in fact alliances and networks of labs working together, to understand the truth—that there are many genes involved in creating something like intelligence, but each gene contributes only a tiny part of the final effect. As Plomin says, for things like intelligence we are looking for a wide sprinkling of "gold dust" (multiple small gene changes), not "nuggets" (rare single gene mutations). The larger the group of people tested, the more accurate the prediction of the sum of these microscopic gene effects.

For the heart, there *are* quite a few "nuggets"—rare single gene mutations that cause certain types of heart disease, such as sudden cardiac death, heart enlargement, or inborn abnormalities of the heart structure. But more common conditions, such as the coronary heart disease that leads to heart attacks, or the arrhythmia of atrial fibrillation, have been tough to pin to a single gene. When we first studied all possible genes from cardiac patients, we hoped to find new insights into the disease. We thought that there might be mutations in a single gene that would explain heart attacks or atrial fibrillation, and that would give us new targets for drugs. However, the results did not tell us much more than we already knew: that things like smoking and diabetes were important risks. Now we real-ize that these common disorders are more like mental disease or weight, where many small genetic influences add up to our lifetime's risk. So now we are scaling up the genetic analysis for heart disease with larger and larger populations to help cal-culate what is called a "polygenic risk score" for each person, which is the sum of the individual small risks produced by many gene changes. Using this score, we can predict a patient's risk more accurately than from any one component alone.[2]

So, the lesson has been learned: bigger is better. Scien-tists, universities, and whole countries are building up groups (called cohorts) of healthy volunteers or patients who are willing to share their information or even submit to special testing. The more people in your cohort, and the more and better information to accompany the genetic data, the more valuable they are. We look at the effects not only of genes, but also environment and lifestyle, on these very large numbers of people. Paradoxically, our vision is that this will let us plan treatments uniquely targeted at an individual patient. Adver-tisers use the same trick to gather data from millions of clicks to send you a pop-up ad for a lawnmower exactly when you were thinking about starting work on the yard.

Precision medicine or "personalized medicine," as this is called, will allow us to get away from the old idea of one-size-fits-all therapy. We have been searching for years to find something that will recognize our individual biologies in all their infinite variety. We know very well that not all patients respond the same way—most drugs in use, even the success-ful ones, work well on only some patients while others con-tinue to suffer. Personalized medicine is a radically different approach: these huge quantities of data have thrown open a door to a new age. This model gives us a way to treat each patient holistically, understanding what factors in their life may have caused the disease and what the best treatment is for them as an individual. Big data is key. But as always, there is a trade-off between quality and quantity.

STRENGTH IN NUMBERS

When we select patients for a clinical trial, we must define very carefully the kind of patient we are treating. They should not only have a particular disease but also be at a precisely known stage of that disease—for example, not just have cardiovascular disease but *persistent* atrial fibrillation or *acute* coronary syndrome or *severe* heart failure, and so forth. We also usually exclude for safety reasons the very old or young, pregnant women, and people who have other diseases as well. Patients who get the active treatment should match the placebo group for numbers of men and women, average age, and the other drugs that they are taking. We do everything we can to get rid of possible biasing factors that could give the drug an unfair advantage over the placebo.

In sharp contrast, the groups that contribute to the large data studies can consist of simply anyone who has gone to a hospital within a certain time period. Researchers can draw on electronic health records from these patients, with the identification of each patient removed. Most countries are now putting everything online and so creating a vast wealth of free data. What these records lack in standardization of the tests, or matching of the patients, or consistency of the treatment, they make up for in the sheer raw power of their numbers.

The journey to the point where we can use electronic health records in this way has been surprisingly hard, given the ease with which we hand out data electronically through our smart devices. The UK's National Health Service (NHS) would seem to be an ideal environment for this, with its birth-to-death tracking of every resident. Other countries with a variety of providers and insurers have more challenges to overcome in harmonizing data systems. But even the NHS has struggled, faced with financial constraints, a variable computing infrastructure across hospitals, and the need for strong data protection. I know that it is not even possible to look at patient information on the university computers in a university-linked hospital medical school because of data protection rules, and UK hospitals are only just abandoning fax machines. Countries that have developed their health systems much more recently, such as China, have immediately used the best technology, so they have an advantage here. But now the problems are starting to be overcome in all countries and the data floodgates are opening worldwide.

A striking example of success from big data has been the predictive power of a simple test for cardiac muscle damage, which is revealed through a rise in blood levels of troponin, a protein normally only found inside cardiomyocytes. Hospitals across the UK contributed electronic health records from more than 250,000 patients, over the period 2010 to 2017. Doctors had been increasingly worried about the troponin test because improvements in the test sensitivity had shown many more people having slight increases. Were these important changes or just day-to-day noise? Troponin is released when heart muscle dies. They knew that 15,000 of these 250,000-plus patients had had a blocked artery, which causes heart muscle to die, as in a heart attack or unstable angina, and the test showed a strong increase in troponin in this group. But they were shocked to find that in remainder of the 250,000, many who came to the hospital for medical problems other than heart disease, a troponin level only slightly raised above normal could predict whether patients would die in the next two years.[3] Doctors have had a wake-up call to take any change in troponin very seriously!

Other unexpected and unexplained findings also came to light. Although people below the age of 40 were tested much less often for troponin generally, a positive test showed they were 10 times more at risk. Very elderly patients (above 90 years of age) naturally have a much higher death rate, but even for them a raised troponin level still showed a 50 percent increase in risk. Surprisingly, most of the deaths in both groups occurred in the first week or so after the test. Doctors would usually have a "wait-and-watch" strategy for these tiny troponin rises, but clearly immediate action was needed. So even this very simple study, with one input (troponin) and one end point (death from any cause), has given some clear instructions to doctors. We don't know what led the clinician to order the test and we don't know what the final cause of death had been, yet big data has given us potentially life-saving information.

THE STORY OF A LIFE

When a patient comes to hospital with heart (or any) disease, the doctor takes what is known as a clinical history. If the patient is elderly, this is very literally a historical record. A 70-year-old would have been born before vaccination programs for rubella, measles, mumps, and polio were widely introduced. They would have been exposed to lead in petrol for around 30 years, and even if they didn't smoke themselves, they would have been exposed to smoking in public places for most of their lives. A 50-year-old man having a heart attack 20 years ago, without fast access to specialized rapid treatments, would have had a much lower chance of surviving compared to today. These large groups, or cohorts, of patients or healthy subjects who have been followed through time are hugely valuable for the wealth of information they give on the life course of disease and the effect of improvements in treatment.

One of the first, most influential, and most enduring cohorts is the entire town of Framingham, Massachusetts.[4] The death of Franklin D. Roosevelt from heart disease in 1945 was the inspiration for its genesis. Half of all deaths in the United States at that time were from heart disease and it was accepted as a natural consequence of aging. Doctors and patients alike were profoundly ignorant about the causes of heart disease. It is sobering to think that no one knew then even how dangerous raised blood pressure can be: hypertension was not thought to be a disease. Doctors accepted blood pressures of 200/100 mmHg as unremarkable for an aging man, while now we know that even 140/90 mmHg is an important cause of ill health, and any rise from the "normal" 120/80 mmHg is considered a warning of hypertension to come. Roosevelt's blood pressure was an astonishing 300/195 just before his death.

Framingham was chosen because it had a population typical of America at that time, mainly white and with a mix of blue- and white-collar workers. It had a democratic town hall system of governance for the citizens to make decisions, and they had previously voted to be part of a tuberculosis (TB) study. Physicians in the town were particularly keen to get involved. Harvard University, that powerhouse of medical science, was in nearby Cambridge and Boston and ready to help design the study and understand the data that emerged. The Framingham Heart Study started in 1948 and managed

to recruit an impressive 5,209 of the town's 10,000 adult citizens, with ages ranging from 28–62 years and, pleasingly,
with almost equal numbers of men and women. It is running
even now, having progressed through four cycles, changing its
makeup of people as the population in the town has changed
and expanded. Children and grandchildren of the original
participants were brought on board, which gave vital information about how heart disease flows down the generations.
When gene sequencing came on the scene, these family trees
were perfectly positioned for exploration of the genetic basis
of heart disease.

A huge amount of our current knowledge started from
the Framingham Heart Study. The researchers drawing from
its data were the first to show many things we now take for
granted: that high blood pressure, smoking, diabetes, or high
cholesterol increase heart disease; that there are differences in
cardiovascular risk between men and women; that exercise,
a moderate body weight, and healthy eating are beneficial.
The phrase "risk score" was first used in a paper describing the
Framingham study and many of the risk factors the authors
used there—age, high blood pressure, high cholesterol, ECG
changes, obesity, anemia, smoking—form the basis for heart
disease prediction today. Other cohorts have confirmed and
added to the findings, from towns in Australia, Wales, the
Netherlands, and Sweden; specific groups such as nurses; and
cohorts targeting age groups, like the elderly. In fact, even the
very latest genetic data in the form of polygenic risk scores is
struggling to improve on the predictions that have come from
studying these people in their daily lives.

New cohorts are always being created, and so new ways of
measuring disease can be added to enhance their value. UK
Biobank has half a million volunteers who were between

40 and 69 years old in 2006–2010; and who have had their genomes sequenced, undergone extensive cardiac and neurological measurements, and provided blood samples for detailed analyses.[5] During the COVID pandemic, the Biobank released critical care data for participants with confirmed COVID-19 to the scientific community in real time. Any scientist worldwide can access UK Biobank data freely on the condition that they are undertaking health-related research that is in the public good. Biobank subjects have agreed to continue to participate throughout their lives, to shed light on the roles of genetics, lifestyle, and environment in the development of diseases of age. Some have agreed to donate their brains for study after death, to help in the fight against dementia. Selfless action by ordinary people is an extraordinary gift to medical science.

LOCATION, LOCATION, LOCATION

Big data tells us that there is lot we can do to help our heart health, but unfortunately there are other dangers that we can't control as individuals. The environments in which we live create hazards that threaten our cardiovascular health every day, and big data is showing us the risks in unprecedented detail. Chief amongst these is pollution. Premature deaths worldwide due to pollution are estimated to be between 8 and 10 million every year, more than war, murder, malnutrition, or road accidents (figure 4.1[6]). We think that at least 50 percent of these deaths are from cardiovascular disease. The two main culprits of air pollution are ozone and fine airborne particles called $PM_{2.5}$, with other agents including the gas NO_2 (nitrogen dioxide) and black carbon particles as strong players. Particles come from petrol and diesel in motor vehicles, but also

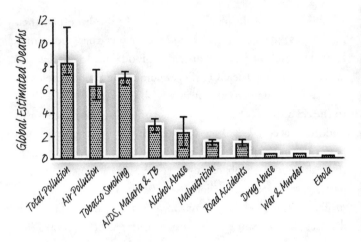

Figure 4.1
Air pollution and mortality: deaths in millions

from natural dust (deserts, wildfires, and volcanos) and inside the home when cooking is done using fuels like wood.

The World Health Organization Air Quality Guidance (AQG) puts a limit of 10 µg/m^3 (micrograms per cubic meter) for PM$_{2.5}$ over a year or a spike of 20 µg/m^3 over a day. This is often exceeded in cities—St Louis in Missouri exceeded 35 PM$_{2.5}$ for 210 days in 2018 and some roads in London usually top the annual AQG limits within five days. In a Medicare population of 61 million people in the United States, an increase over the previous year of 10 µg/m^3 for PM$_{2.5}$ increased death rates by more than 7 percent.[7] Even in healthy people, UK Biobank found that higher past exposure to PM$_{2.5}$ was associated with cardiac enlargement, a marker that often precedes development of heart failure.

Walk for just two hours in certain parts of London and you can start to damage your heart. A study in the city showed dramatically the real damage of air pollution in everyday life.

Healthy volunteers and people with stable lung or heart disease were assigned randomly to walk at their chosen pace in Oxford Street, a famous shopping area, and at another time in nearby leafy Hyde Park.[8] They carried equipment to measure pollution and noise levels with other information coming from sensors along the road. All the people who walked in the park had increased lung capacity and measures of blood vessel health, and these benefits of even such gentle exercise lasted for more than a day. But when they walked in Oxford Street, with its stream of diesel taxis and buses, the effect was quite different. Walking here produced no health benefit and, even for disease-free volunteers, there were clear signs of blood vessel harm. (Cyclists tested with diesel exhaust in the laboratory also had disturbing changes in lung function, suggesting too that riding along busy roads could cancel out some of the health benefits of exercise.) The difference in $PM_{2.5}$ between the two walks was not all that great: Hyde Park averaged just below the 10 $\mu g/m^3$ limit and Oxford Street just over the 20 $\mu g/m^3$ one, yet this was enough to produce the shocking contrast in effects on health.

In a piece of good news, people on statins and blood pressure medication in the heart disease group were relatively protected against the harmful effect of pollution, and this gives a clue that the pollutants are driving forward the disease process itself. We knew that gases like NO_2 could penetrate the body, but now scientists have discovered that even $PM_{2.5}$ particles can enter through the lungs and reach tissues such as those of heart and brain. Both the gases and the particles have a host of harmful effects, stoking the fire of inflammation and dangerously increasing the chances of developing diabetes.

The walking study also showed that noise levels were a further factor in the harmful environment. Both healthy volunteers and the heart disease group showed a decrease in blood vessel health related to noise levels in Oxford Street. Airplane noise is known to increase blood pressure in those who live under busy flight paths, with sleep disturbances being a key factor. Harm from the environment is clearly much wider than air pollution alone. Urban design itself can be damaging to the health of inhabitants, when it brings main roads toward vulnerable areas such as schools; reduces green space for recreation; and discourages walking and makes people rely on cars for transport.

What can we do about these worrying statistics? Personally, we can avoid polluted areas for prolonged periods—there are apps that track road sensor information—and buy air conditioners and air filters for the home. But we all need to work together not only to change our individual choices but also to influence public policy. In many ways the health agenda converges with that of combating climate change: global reduction of fossil fuel emissions and other greenhouse gases, changing to a healthy plant-based diet, reduced reliance on driving and enhancing the urban and rural environments. Imaginative and large-scale public health policies, like the previous smoking bans and prioritizing of emergency care, are once again key to preserving cardiac health.

NATURAL LIVING—HUMANS IN THE WILD

In my latest laboratory we have an apartment in the middle of the floor, all set up with a bed, couch, TV, kitchen, and bathroom. Not, sadly, intended for tired scientists to take a break, but to study how people will interact with new technologies for remote monitoring and care at home. With them we can extend hugely the Framingham model of study, where so much has been learned from people living ordinary lives in their own towns. New technologies will also free patients from the depressing grind of frequent hospital admissions or appointments, with all the travel and disruption that comes with them. In the new post-COVID era, these technologies are coming into their own, having been accelerated by the urgent need to control the virus.

Many of us are now comfortable with our smartphones and smartwatches that track our step count, stairs climbed, heart rate, and even ECG. We love the positive feedback given by knowing our day-to-day achievements—this incentive can even become an obsession. Scientists are making rapid headway in developing algorithms that can use rules to interpret ECGs, for example, and alert when a dangerous arrhythmia is detected. Technologists at the same time are developing less intrusive and more user-friendly hardware to gather information discreetly. Earphones can measure heart rate from sensing blood flow changes or the ECG from electrical changes on the surface of the skin.[9] Wearables can record important predictors of wellness such as speed of walking, or rising from a sitting position, using the same accelerometers that measure step count in smartphones. In-ear sensors could also be extended to measure breathing, from the turbulent air flow in the ear canal, and this would replace the cumbersome and restrictive chest bands now used for out-of-hospital monitoring. Even brain activity, usually gathered by experts in a specialist laboratory with highly intrusive electrode nets attached to the skull under the hair, could potentially be sensed through an over-ear attachment. The more discreet and unobtrusive the device, the less it will disturb normal activities when used.

There are challenges of course—if you have a device in your ear, then when you speak or eat the noise will interfere with the signal. But even this waste "noise" could be harnessed. Scientists are using these chewing and swallowing patterns to find out what sort of food you are eating—notoriously difficult information to obtain from self-reporting—or whether you are taking your pills. As you speak, the patterns of highs and lows in your voice, even stripped of the meaning of the words spoken, can give valuable information about you. The pitch, jitter, energy, rate, or length and number of pauses can indicate the moment-by-moment levels of stress that you are feeling.[10] Researchers think that your blood pressure could even be measured remotely from the ear sensor or from a telephone conversation using these frequency variations in speech.

Placing devices on the wall of a room or building, to measure our body processes without physical contact, are now also becoming routine. We have seen the heat scans of passengers walking through an airport arrivals hall, when a new epidemic emergency like COVID-19 makes it urgent that we identify infected people entering the country. Miniscule fluctuations of this heat pattern can also be used to detect heart rate—maybe in the airport to identify an infected passenger, or in a home setting to detect the arrhythmia of a heart disease sufferer.[11] New sensors using a combination of radar and radio-frequency waves can measure blood pressure from monitors in the corner of a living room.

Imagine all these new sensors installed in the house of your elderly father with heart disease. You can see when he wakes up, and whether he struggles to get out of bed—you can wish him good morning through a voice or video link. Immediately you have a full status check on his temperature, heart rate, and blood pressure. You know that if his heart shows a disturbed rhythm or the sensor shows he has had a fall, both you and the dedicated nurse will be alerted immediately. Possibly you could see whether he has had a meal or drink and is taking his medication. You can stock up the smart fridge or order a prescription accordingly. The hospital shares this record of daily living and has a real-time appreciation of his physical health, mood, activity, and general quality of life. He has his valued independence but is cared for without fuss by invisible hands.

Think, however, of the volume of data that even one person would generate living like this, with thousands of data points per minute over 10 years of life. In Framingham, each of the approximately 5,000 patients might have had 30 measurements per year for 40 years. In UK Biobank, as well as the detailed biochemical and imaging data (100,000 cardiac scans for example) for the 500,000 subjects, there are whole-genome analyses. The 32 billion base-pair information from each person is equal to the number of letters in a book more than 500 times the longest novel ever written. Bigger still is the volume of potentially useful health data that most people give away daily, knowingly or not, through their smartphone and social media. Where you walk, for how long and how fast, and in which environment can be linked to maps of urban pollution and crime, and to local incidence of disease. What you buy to eat can be recorded and what motivates you to buy healthy or less-healthy foods can be deduced. Your social connectivity too—a strong indicator of mental health—is easily calculated and given a score. The untapped potential for understanding, and ultimately influencing, heath-related behavior is huge.

DEEP DATA—A UNIVERSE WITHIN A CELL

While the population scientists have been going wide, the laboratory scientists have been going deep and this has sparked another revolution in how we understand the body. When

I started in the laboratory, all the biological research we did used the "hypothesis-driven" approach—that is, you had an idea about a molecule that might be involved in heart failure and tested it by a series of experiments specifically investigating that molecule. You might use a drug to stimulate or block it, or you might increase/decrease its levels in a cell or in a mouse by manipulating the genes. Either you can cut out the gene itself (gene editing) or stop it working to produce its final product, the protein. These detailed experiments are still vitally important when we get close to the final goal, but what big data has done is vastly expand the number of possible candidate molecules at the start. This is the "hypothesis-generating approach."

Today, we don't just look at how one molecule changes, we also see in a holistic way how all the molecules change at once. We have gone from looking out the window to see if it's raining to mapping global weather patterns. We can see how a hurricane in Florida will produce a storm in France. In the same way we can see how a hormone binding to a receptor molecule on the outside of a cell causes a cascade of unsuspected changes on thousands of proteins. A whole organ or organism adapts rapidly to any stimulus by activating or reducing many molecular pathways at once. Measuring just one protein at a time could never explain the complexity of the biological response. We now think about biology in a "systems" way. Instead of describing the effect on one protein we talk about clusters of molecules that have a general effect—cell division, growth, death. The biologists rely on advanced mathematicians and statisticians to analyze these huge amounts of data, and it has become a science in itself, called bioinformatics. We have devised ways to explore the data without the bias of our previous knowledge, by letting clusters of meaning emerge unsupervised from the analysis: this has yielded astonishing new insights into disease.

There are five billion cells in the left ventricle of a human heart: cardiomyocytes make up the vast bulk of the tissue, though they account for only around half of these cells. When we look at the whole heart, we are seeing an orchestrated sum of the actions of many cells: not only the cardiomyocytes but the cells of the blood vessel (endothelial, smooth muscle, pericytes) and the supporting cells like fibroblasts and resident macrophages. As we zoom in from above, we can see the heart as a city, with the arteries and veins as the highways between the solid office buildings that are the cardiomyocytes. These buildings cannot survive alone but need the sustaining presence of shops, restaurants, police stations, and all the services that a city brings. In the same way the cardiomyocytes need their attendant cells for support, sustenance, and repair.

Location is everything. In the heart too there are different parts of town with very individual characters and needs. The blood vessel lining of the tiny capillaries is not the same as that of the aorta, the great vessel carrying blood from the heart into the body. Communication is key, and the flow of people and traffic between buildings, as well as the invisible communication links of the electronic world, are paralleled in the multitude of messages that stream constantly between the cells.

When danger threatens, this communication and coordination can make a city function as one to respond to events. In the grip of a pandemic or a terrorist attack the systems will switch from normal day-to-day activity of many individuals to a synchronized reaction—lockdown, mass evacuation. When we analyze all the gene products in many individual cells within a tissue, we see that they are often just doing their own thing. Each cell may be making the same protein but with different efficiencies—some very actively and some almost dormant. Over time, every cell will be going through a cycle of low to high protein manufacture and back. But when there is a threat and this protein is needed, every cell will ramp up to the maximum production rate in an instant. Because they were all active already, they are quick off the starting block: this system is much more efficient than every cell starting from scratch.

Understanding this variation between individual cells is so important that institutes like the Broad Institute in the United States and Wellcome Sanger Institute in the UK have created an exciting international project called the Human Cell Atlas.[12] This is the new "moonshot" of biology because it aims to analyze not only what every cell is doing, but also the exact address of that cell within an organ. But when we multiply the number of cells in the body by the number of active genes in each cell, and we realize that the activity is changing second by second, then the imagination struggles to understand how we can encompass this complexity. To paraphrase J. B. S. Haldane, the universe may not only be more complicated than we think, but also more complicated than we *can* think!

So we have wide data and deep data, and both are expanding exponentially as we develop newer technologies. Looking at the graph of health data (figure 4.2) we can see the overwhelming increase over only a few years. Yet the number of scientists to analyze the data is not growing much at all. We definitely need some help!

DATA DELUGE AND THE ARTIFICIAL MIND

One thing our brains are really good at is recognizing patterns. Every civilization has seen its mythology written in the star

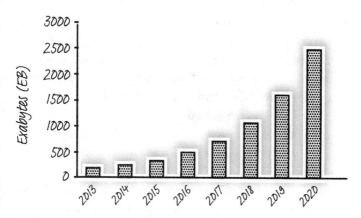

Figure 4.2
Increase in health care data from 2013 to 2020. One exabyte is about equivalent to streaming one billion movies.

patterns in the sky and imagined faces in the rocks and clouds. We need not get too proud of ourselves—bees and pigeons are probably better at navigating by remembered landmarks than we are. This is a key evolutionary aspect of the biological brain. For every clinical diagnosis we make, we take multiple complex images and measurements from each patient, so the ability to recognize patterns is a huge part of the health care data challenge. Our urgent goal is to teach the computers to take over these tasks—to harness the power of artificial intelligence.

Pathologists, the people who look at the microscope slides to see the signatures of cardiomyopathy, or radiologists, who look at heart scans to pick up the hallmarks of scar, are pattern gurus. Many years of experience have given them an almost preternatural instinct for the faulty cell or anomalous heart movement. Now they needed to find a way to hand this task over to the computers. The entry portal from human knowledge to AI reading of scans is through supervised machine learning. First, a "ground truth" or agreed real-world result, must be established through many thousands of examples of scans that have been looked at by several experts. They must agree on the best line to draw on the image to measure the complex archipelago of the inner heart wall or what faint deformity of a valve signals its first decline toward failure. Then they must hand this library of wisdom over to the machines.

Artificial neural networks, which mimic the action of neurons in the brain, are computing systems with interconnected nodes that use algorithms to recognize hidden patterns in large datasets. The complex computer network absorbs the vast repository of knowledge from the human-analyzed scans and then sets about trying to reproduce the results. In each round, the network assesses its own performance. The number of features of disease that it has missed (false negative) and the number of normal features it has mistaken for diseased (false positive) are calculated and it gives itself an accuracy score. After each cycle with the training dataset, the output from the machine learning code is compared statistically with the ground truth. The program cycles back again with this information and makes improvements. The algorithms within the computer code give it the ability to automatically learn and adjust from experience without being explicitly programmed. When the accuracy score reaches 80 percent or above, the machine learning algorithm is approaching something useful, close to an expert operator.

The cancer field has seen the best success with this, with an AI program developed that is better than expert radiologists at spotting breast cancer in mammograms. Progress has been made in cardiology: a machine learning algorithm trained on

90,000 heart images marked up by experts is now able to handle new images with a performance equal to an average human observer.[13] The advantage is that the algorithm can do this in three seconds, compared to several minutes for the expert, and it can work 24/7 without getting tired or distracted. We can also use machine learning to estimate maximum blood flow through a partially blocked blood vessel, which is important to predict whether the blockage is likely to result in the patient experiencing the chest pain of angina. But the movement of the heart adds another layer of complexity compared to a static mammogram.

Moving MRI, known as cine-magnetic resonance imaging (cine-MRI) of the heart, records a wealth of hidden information, but we usually take from cine-MRIs only a few measurements of size and shape. So, we look to the artificial intelligence researchers to help us detect patterns we would not be able to see ourselves, to suggest new hypotheses that are hidden within this mountain of data. Here we turn to something much closer to true artificial intelligence, called *deep learning*, which uses layer upon layer of computational networks to remove the need for the supervised learning of training datasets. Inspired by the architecture of the brain, these "neural networks" can work unsupervised to detect their own patterns and present them to us. When scientists were teaching computers to play the ancient and complex game of Go, with its possible moves outnumbering atoms in the known universe, they first used supervised training from hundreds of games already played. This was moderately successful, with a computer eventually beating a human player. But with another program they just gave the computer the rules of the game and let the computer play against itself unsupervised. Very quickly it came up with moves that no one had ever seen before, and which were without apparent logic, but nevertheless this program was able to beat the human players time and time again.

Researchers set their unsupervised neural network to solve a problem that had them puzzled. They had shown that the widespread mutation in the cardiac protein titin, which was found in a high number of patients with dilated cardiomyopathy, was also present in a significant number of apparently healthy volunteers. The usual imaging tests were not able to detect any signs of dilated cardiomyopathy in these people, so the program was set to see if it could find anything different in their hearts. By itself, the program was able to discover a subtle bulging of the borders of the heart in the healthy-titin carriers, a possible precursor of damaging left ventricular remodeling.[14]

In another study, AI was even able to help predict death. Pulmonary hypertension is a severe and life-threatening disease where the right side of the heart is overstressed by obstructions to blood flow through the lungs. AI was able to pull out signals from the heart scans of 256 patients that more accurately predicted their likelihood of death within a year than normal imaging or functional and clinical markers alone.[15]

These are early studies, but they begin to show the power of the AI methods, giving a new superpowered set of eyes to researchers viewing the complexity of heart motion. But as with much of the AI generated, there is a dark side. Privacy is becoming more of a notional concept generally as our own data is distributed through our use of mobile devices. Huge efforts have gone into protecting the anonymity of health data because of their sensitivity, but we may not be able to hold back the tide. Your smartphone is quickly able to deduce without being told that your daily journeys are between work and home: those two pieces of information will immediately allow you to be identified with high accuracy. Further, we must be careful not to contaminate AI with our own biases by feeding in data that will make the programs reproduce our prejudices. Amazon found that its AI system for sorting job applications was routinely rejecting those from women.[16] The historical records on which it was trained had lacked women, so the program absorbed this as a rule to take forward (and was in fact reinforcing it more strongly over time). Training data-sets for health data must be balanced for sex, age, ethnicity, and other important characteristics.

Even the uncanny power of AI to pull unsuspected patterns from the data has its flip side. If we don't know how the result was arrived at, how can we understand what the program was looking for? How can we be sure that mistakes or biases were not there? When the AlphaGo or Cardiac AI program gives us an unexpected new result, we need to go back and untangle what led it to that conclusion. As in *The Hitchhiker's Guide to the Galaxy* by Douglas Adams, we will not be satisfied with knowing that the "Answer to the Ultimate Question of Life, the Universe, and Everything" is 42—we need to know what the question was.

THE PLASTIC HEART

The heart is a remarkably resilient organ. When you consider the number of insults it suffers, old ones from the dawn of time and new ones we have only just invented, the wonder is that it keeps going as long as it does. Many more people are living into healthy old age now and operations to patch up faulty valves or implant pacemakers are done on 90-year-olds and older. Drug trials confirm that taking statins keeps improving heart health in people of 75 years and up.[1] This resilience is even more impressive when you consider that the heart has such a limited ability to regenerate itself by producing new muscle cells. A few years ago, we thought that there was no regeneration at all. Only a freak accident of pollution gave us a definitive answer, by allowing us to carbon date the heart.

Carbon dating is usually associated with archaeology, and you may have heard about it in connection with fossils, dinosaurs, or ancient artifacts. It measures the amount of radioactive carbon (carbon 14) in a tree or animal. Most of the carbon in our bodies (and the bodies of animals and trees) is carbon 12, meaning that it has 12 protons in its nucleus. We continually take in minute traces of carbon 14, which is produced by cosmic rays hitting carbon 12 in the atmosphere and adding extra unstable protons. When an organism dies, the amount of carbon 14 drops with time and is not replaced. So the carbon 12 to carbon 14 ratio tells us how long ago the organism last took in carbon 14, and therefore when it died. This is a very small change and works best over the timescale of hundreds or thousands of years.

However, in a unique natural experiment, overground bomb testing in the 1950s and 1960s suddenly created a spike of radioactive carbon in the atmosphere, and this was taken up into bodies and cells of living people. Then someone realized this was undesirable (to say the least) and that it might be a good idea to conduct the testing underground. The level of carbon 14 in the atmosphere began to drop with time. Cells in the body of someone living around this time that had been created after the person was born then had lower levels in their nucleus than those cells that had been present at birth. When someone died, scientists could measure the carbon 14 in all their cells and see which they had had since birth, and which were created more recently.[2]

Amazingly, about half of the cardiomyocytes in the heart of someone aged around 75 had been present for their whole life. Each individual cardiac muscle cell, less that one tenth of a millimeter long, would have been beating for 75 years continuously: more than two billion contraction and relaxation cycles. This is perfection on an almost unbelievable scale. There was a tiny smidgeon of regeneration too, however: about 1 percent of cardiomyocytes were renewed in a year for a young person, dropping to half that value in a 75-year-old. This could be just enough to cover you for loss due to general wear and tear, the daily grind of anxiety and effort, starvation, and natural insults. But it was nowhere near enough for the train wreck of severe cardiac disease—the body's defenses are completely overwhelmed by the massive destruction of a heart attack.

The brain scientists too have come through the same journey to understand if there is any regeneration in the neurons of the central nervous system. Again, the field has gone backward and forward between complete denial that any new neurons appear after birth and the idea that there might be few, or in localized areas or only after there has been an injury. Again, it was carbon dating after the bomb tests that managed to pinpoint the tiniest of signals that showed some trace of new cells.[3] But clearly the brain is hugely adaptable—that is the basis of our whole evolutionary success and the blossoming of civilization. This adaptability is not from growing new neurons, however, but because of the staggering plasticity that comes from the continual formation and breaking of neuronal connections—the synapses. We can instantly adapt to experience over a very short timescale, such as when we pick up an unfamiliar tool, or more slowly over months and years as when we learn a new language. We can repurpose different parts of the brain when one area is injured: even into old age there can be dramatic recovery from a stroke. This is neuroplasticity, the driving force for the malleability of brain function.

Plasticity is also the core of the heart's ability to adapt without regeneration. It can change massively in size in a relatively short time—a few days or weeks—when the demands of the body alter. The heart can double its volume in pregnancy or thicken its wall by 50 percent if the aorta carrying blood out of the heart is narrowed. The heart wall of an elite athlete can be 20–30 percent thicker than normal without any sign of disease. After damage, the cardiac wall can thin and balloon out—by greatly expanding its volume, the heart can magnify the effect of even a feeble beat.

A human-made experiment showing the extreme plasticity of the organ is the "man with two hearts" operation, or heterotopic transplantation. In the earlier days of heart transplantation, the donor/recipient match was not always good. When only a small heart was available for a large donor, surgeons would plumb in the second heart without removing the first. Usually, they would put it some way down the aorta in the stomach region. Stories abound of teasing medical students by asking them to find the heartbeat without telling them there would be two! These secondary hearts would add something to the flow of blood, but without the stimulus that the first heart receives they would dwindle in size.[4] When it is not doing normal work, the second heart can reduce in size (atrophy) by about 30–40 percent within 5 to 10 days. The partial artificial heart assist devices, which take over some of the blood flow, have the same effect of reducing heart size rapidly.[5]

ATHLETES VS. THE REST OF US

When demand increases the opposite effect occurs, and the heart responds by growing its muscle layer. This seems a simple idea, just like when you exercise your arms by lifting weights. This demand can come from exercise and, in our evolutionary past, this is the main way the heart was stressed—hunting prey for many miles, carrying children over long journeys, gathering enough plants and fruit for a large tribe. Exercise is absolutely vital for our heart health and the response of the cardiac muscle is well adapted to normal or even strenuous activity.

Building up the muscle layer of the heart does not result from an increase in the number of cardiomyocytes, but from the change in their shape. These individual cells can increase or decrease their width in response to the mechanical stress they feel, bulking up when they are working hard and slimming down when the load is reduced (figure 5.1).

To perform the work they need to do, cardiomyocytes are packed with two kinds of fibers: myosin and actin. Myosin and actin slide over each other—or to be more precise, the

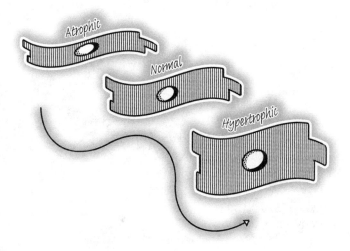

Figure 5.1
Changing cardiomyocyte size from low (atrophic) to high
(hypertrophic) workloads

myosin crawls up the actin by sticking and unsticking its protein heads. Since both myosin and actin are attached to the
ends of the cell, this movement pulls the ends together and
makes the cell shorter. It is this shortening of individual cardiomyocytes that powers the contraction of the heart. When
cardiomyocytes sense a continuous increase in load for a long
time (days or weeks) they increase the number of myosin and
actin molecules and get longer or thicker (hypertrophy). In
turn, that makes the heart wall get larger and increases the
amount of blood ejected with each beat (figure 5.2).

EXTREME ATHLETES—TOO MUCH OF A GOOD THING?

Exercise is good for you, there is no doubt about it. But we
humans are always taking things to extremes, and there is a
point where even exercise can turn bad. Endurance athletes

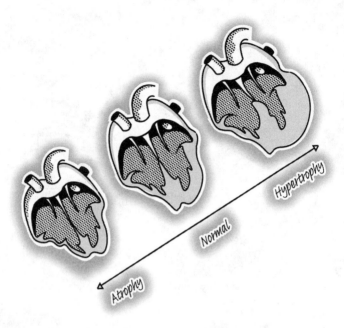

Figure 5.2
Changing left ventricle shape size during atrophy (low workload) to
hypertrophy (high workload)

are a breed apart, performing feats of astonishing strength and
stamina. The record at the time of writing is 59 successive
days running a marathon, but no doubt this has been broken
since. Interestingly, endurance cycling is one of the toughest
sports for your heart, probably because of the combination of
large leg muscles needing steady maximum blood flow and
the intermittent peaks of intense work for the arms and upper
body. Top blood pressures of 200 mgHg are common during
rapid mountain climbs as in the Tour de France. The cyclists
must also keep this up for long periods—up to five hours—
compared to the two to three hours for a marathon.

Overall, endurance cyclists do well in general health, with mortality reduced by 41 percent and life extended by 17 percent.[6] But a significant proportion do develop heart abnormalities, about 50 percent more than in the general population. These include a wide range of arrhythmia types including atrial fibrillation. A thick (or hypertrophied) heart wall is much more prone to rhythm abnormalities. When 46 athletes (37 cyclists) who had these abnormalities were followed up over five years in one study, 18 had a serious cardiac event. Nine of them (all cyclists) died within two years. We must always remember that evolution balances the increased risk of overstimulating the heart with the benefits for immediate survival in times of danger. Extreme exercise saves lives in an emergency but is risky as a modern lifestyle choice!

AN UPHILL STRUGGLE

The advice to avoid extreme exercise is all too easy to take. But we can unknowingly give our hearts a dangerous burden that it carries every day. High blood pressure, or hypertension, means that your heart must push against a greater load every time it beats. Probably you have seen the blood pressure readings from your clinician and are familiar with "120 over 80" as the ideal. The higher figure in your blood pressure reading is the pressure of the blood flow at the peak of the heartbeat (or systole). It is measured, quaintly, in millimeters of mercury (mm Hg) as that was the height of a column of mercury in the old sphygmomanometers (blood pressure machines) that was needed to produce the pressure to inflate the cuff and cut off the blood flow to your arm. A high systolic blood pressure is immediately dangerous as it can trigger such things as brain hemorrhages or strokes, and in the long term contribute to heart and kidney disease. A blood pressure of over 180 mmHg at rest is the trigger for an emergency hospital admission.

The lower number on the 120/80 is the pressure when the heart relaxes (diastole). Changes in diastolic blood pressure are not usually so dramatic, but they are important. This is the pressure that your heart must work against, 100,000 times a day, day in and day out. Even an increase from 80 to 90 mmHg is putting your heart under increasing strain. Again, your cardiac muscle will bulk up and thicken to match this. However, this is not the benign strengthening that comes with intermittent exercise, or physiological hypertrophy, but a whole different and dangerous adaptation—pathological hypertrophy. Here again, the body plunges into the spiral of destruction that triggers heart failure, where hormones and neurotransmitters are activated mistakenly by the body to counteract a perceived threat from early evolutionary times. In fact, untreated hypertension alone can cause heart failure and the armory of drugs for high blood pressure overlaps significantly with those used to treat heart failure. In the UK there are 15 million people with high blood pressure and in the United States 108 million—among the over-50-year-olds this may be half the population.[7] Many may be untreated or not fully treated. High blood pressure in this context is often defined as greater than 140/90 mmHg, but the threshold is moving to 130/80 mmHg.[8] About half of all heart attacks and strokes are associated with this disease.

Lifestyle changes can protect us to some extent: decreasing salt in food, lowering alcohol intake, eating a generally healthy diet, losing weight, and avoiding stress. And most important, don't get old! However, until we all absorb the good advice and return to our ancestral ways, the life of an average worker in a developed country will act against any good intentions.

Blood pressure medication will certainly be needed. Many people will not be treated fully with one blood pressure drug, but may need two, three, or even four. Even then, cases of resistant hypertension are increasing, and surgical solutions are being developed. People of African descent are particularly susceptible and may need more and different kinds of antihypertensives.[9] Several hundreds of billions of dollars are spent in the United States alone every year on these drugs, and this amount will rise with an aging population. Clearly, this is a challenge we have not yet resolved. We need some new science to think more deeply about the long-term insidious damage to our hearts from increasing workloads.

PATTERNS OF DISEASE

Why is increasing the workload of the heart with intermittent exercise good for the heart, but doing the same by a slight diastolic blood pressure increase is bad? Pathological hypertrophy looks like physiological hypertrophy from the outside, so why is it so much worse in the long term? To investigate this, scientists first started with the "molecular biology" approach, manipulating individual genes and proteins in our laboratory models. This is how we found the angiotensin inhibitors for both heart failure and hypertension, now some of the most widely used drugs. But when the gene detection technology got cheap enough, researchers could change to the "systems" approach. When we stressed our models to produce hypertrophy, were able to understand how all the gene messages for all the proteins in the heart changed at once. We could see a cascade of effects radiating out from the initial increase in workload and up- or down-regulating thousands of genes with a pattern evolving over hours to weeks.

When we did this for pathological hypertrophy, we could see that there was a completely different pattern of gene changes emerging than for physiological hypertrophy. Amazingly, the heart seemed to be regressing in time, going back to its earliest origins in the womb. As the heart becomes adult, some proteins subtly shift their structure to perform differently. The myosin protein of the muscle fiber, for example, becomes able to work more quickly to give the mature function of the human heart.[10] We could see these gene and protein changes reversing in the pathologically hypertrophied heart and the muscle, essentially, returning to factory settings. We think that this is a protective mechanism, as the body senses a mismatch between workload and oxygen supply in the thickened muscle. It is returning to the pattern in the embryo, which must survive in the hypoxic conditions of the womb. The heart muscle becomes more efficient to save using oxygen, but this is done by reducing its speed of contraction and, crucially, of relaxation. Once again, the poor relaxation of the heart is the culprit in disease.

IT TAKES A VILLAGE

The story of hypertension gives us another very valuable lesson: the cardiac muscle and blood vessels are an intimately connected system where one cannot adapt without the other. More than this, the cardiomyocyte is far from being the only player in the heart wall: while it is the largest contributor by volume, it is outnumbered by other (smaller) cell types. In addition to the cells making up the blood vessels there are immune system cells and structural cells like fibroblasts supporting them all. It's a community of cells, and this community is constantly exchanging information. When we look through time-lapse cameras at cells in a Petri dish, we can see tiny tentacles reach out to move the cell around like legs, or to probe other nearby cells like fingers. I have a great video that shows two different fibroblasts (scaffolding cells) meeting a cardiomyocyte for the first time. Fibroblast 1 is very curious. It rushes into the shot and encounters the cardiomyocyte, which is not moving about but just sitting there quietly beating. The fibroblast puts out feelers to touch the cardiomyocyte surface and even seems to nibble off some of the contents. It happily caresses the whole surface of the cardiomyocyte and then gradually departs. In a neighboring dish, fibroblast 2 is not so impressed. It barely touches the cardiomyocyte before it withdraws hastily and scrunches up, a flood of ripples disturbing its surface. It departs with extreme swiftness as if fleeing from a dangerous foe. Both fibroblasts have sensed something about their respective cardiomyocytes through the physical contact between their two outer membranes. Did fibroblast 1 meet a healthy cardiomyocyte whereas fibroblast 2 found a defective one? Or maybe fibroblast 2 was just having a bad day. At the moment, we have barely begun to decode these messages.

Without even touching, cells and organs continually message each other—both their next-door neighbours and those on the far-flung continents of the body. We have known for many years that hormones are released from one organ to influence others—for example, angiotensin II is released from the lungs to control water balance at the kidneys: this is the endocrine system. Recently we have learned more about the paracrine system where individual cells can contact each other by releasing similar hormonal factors—like a local gossip network. Now we realize that these messages can be sent out as packages of information, rather than a single factor. If we look at the surface of the cell, we can see lively release of tiny buds

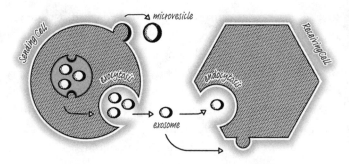

Figure 5.3
Communication between cells by export of cell contents in
microvesicles or exosomes. The receiving cell can import the message
by direct fusion of the exosome or vesicle to the outer membrane or
endocytosis (engulfing it).

of membrane, which float out into the surrounding fluid like
bubbles. Inside these bubbles are packages of many proteins
and microRNAs, giving subtle and complex suites of instructions to the lucky recipients of each parcel. Smaller packages
called exosomes are assembled inside the cell and released in
a burst, while larger ones are called microvesicles and bud
directly off the outer membrane of the cell (figure 5.3).[11]
When they reach their destination, they may get swallowed
up into the receiving cell (endocytosis) or bind to receptors on
the outside.[12]

We know some things about this process, but many remain
mysterious. We know that the messenger molecules can have
a multitude of effects on the cells they encounter, to stimulate and inhibit; preserve or kill. Beneficial effects on the heart
from exosomes have been shown many times in laboratory
experiments. We know that disease alters the content of the
packages—exosomes from the fluid surrounding the heart
are normally beneficial, but some from patients with cardiac

disease can actively harm the tissue.[13] But we do not know the intention of these exosomes or microvesicles. How specific are the package contents—are they tailored for different situations and different destinations? Are they intended for only one receiver cell type, with a precise address label? Or more like a general signal to all cells? Even without the answers to all these questions some scientists are thinking about clever ways to use these exosomes to deliver cargo/therapies we might want to send to cardiomyocytes.[14]

The mechanisms of plasticity can be fast, on a scale of hours and days. Deployment of different genes is controlled by how quickly their messages are translated into new proteins, and this starts within minutes of a new stimulus. The protein levels will begin to alter immediately, and significant changes can be measured within hours, but changing the structure of the heart takes longer, more like days. The function of proteins can also be turned up or down more rapidly, by addition of chemical groups such as phosphate, for increased flexibility. However, there are responses which need to be considerably more rapid than that. In evolutionary times, survival depended crucially on how fast your heart could respond to an immediate threat to life. For this, there is a whole separate nervous system that brings the timescale of response down to minutes, seconds, and milliseconds, and operates without your conscious control.

THE RESPONSIVE HEART—EMOTION IN MOTION

The beating of the heart is the rhythm of our daily life. Every time we wake and rise from bed, lift a sleeping child, or run for the bus, the heart is constantly adjusting to support our needs. Second by second and minute by minute our heart is adapting its beating rate and force to match the movement of blood to the muscles, skin, and brain. The heart is responding to our emotions but unexpectedly, as we have only just discovered, creating them as well. The mechanisms that control this are unbelievably ancient and their precise honing over millions of years of evolution has been crucial to our survival as individuals and as a species. They are too important to be left to your voluntary supervision and so they operate silently, day and night.

YOUR SECRET CONTROL NETWORK

You can't live without your heart, but your heart can live without you. One of the most unnerving sights is that of a human heart, completely removed from the human body in the operating theatre, beating slowly by itself. I have seen this when watching the surgeon transplant a new donor heart into a patient with end-stage heart failure. I'm usually there waiting for them to dissect a sample of the diseased and discarded heart to give to me for my experiments. The apparent life remaining when the heart is removed shows powerfully that it can beat without conscious control—luckily, or we would never be able to sleep! So we don't need a mental effort to drive every beat, but we do need a way to fine-tune the heart output for times of emergency or times of rest.

In the body, heart rate and contraction force are controlled to an exquisitely fine degree by the actions of the autonomic (as in autonomous, self-governing) nervous system. Many other processes are also under autonomic control: blood flow, digestion, and breathing for example, and they too carry on perfectly without our conscious intervention. There are two separate nerve networks in the autonomic nervous system— the sympathetic and parasympathetic systems—and this pair of systems provides the accelerator and brake for the heart.

The sympathetic nervous system is the accelerator, preparing us for action, taking us out of danger. Even the name with its link to sympathy gives the clue to the intense involvement with our emotions. This is the classic fight-or-flight response that evolved with us in the time of saber-toothed tigers, and where adrenaline (with its close relative noradrenaline) is the key. When preparing for conflict, adrenaline and noradrenaline are released and blood is pumped away from the skin and gut and into the skeletal muscle of the legs, preparing for quick action. Digesting your lunch can wait while you make sure you don't end up as something else's! Your face is white from the bloodless skin and there are butterflies in your gut from the knotted muscles. Your heart pumps harder and faster, but also the total contraction and relaxation with each beat happens in a shorter time. This is to allow the relaxed phase of the heart to take up as much of the time as possible to fill with blood. Otherwise, the output of the heart would not be able to keep up at the faster rates.

The parasympathetic nervous system runs parallel to the sympathetic and opposes it in every way—it is the dimmer switch, the calming pause, the yin to the yang. Your heart slows to a steady rhythm and contracts less strongly; the blood redistributes to the restorative functions of digestion and secretion. Your lunch resumes its orderly transit. Surprisingly, your sympathetic and parasympathetic nervous systems are operating at the same time—the body keeps the accelerator and brake on together for a quick getaway or rapid relaxation. If both were suddenly blocked then, strangely, your pulse rate would go up since the parasympathetic system contributes more to your resting cardiac heart rate. Therefore the crash team in the emergency room is yelling for atropine (the parasympathetic system blocker) as well as adrenaline (the sympathetic system stimulant) to revive the patient in cardiac arrest.

We can see how important both systems are by the number of fallbacks and fail-safes in the sympathetic/parasympathetic crosstalk. The two systems' interaction is intricately controlled from the most high-level point, where neurons in the brain link across to allow one system to inhibit the other, right down to the molecular level where multiple proteins within each cardiac cell perform opposite functions for each system. Neurons that sense the pressure in the carotid artery in your neck—the baroreceptors—monitor your blood pressure from second to second and fire more quickly as blood pressure rises. The parasympathetic system then responds by damping it down—lowering your heart rate and increasing the flow through blood vessels. When baroreceptor firing is low, sympathetic activation increases the heart rate and blood vessel resistance to bring the heart back to an ideal point for the activity at that moment.

The dynamic tension between parasympathetic and sympathetic systems causes minute fluctuations in cardiac output—for example the time between heartbeats varies on a millisecond scale from one to the next. Although you might think that heart rate variability could be a negative thing—surely you want a steady heartbeat?—on the microscale it shows that both systems are working together to produce this exquisitely fine-tuning. Heart rate variability can be analysed from a portable ECG monitor (or now, a wristband device) that will show the health of your heart control mechanisms. Poor heart rate microvariations are an important warning signal for patients whose hearts are starting to fail.

YOUR HEART CAN TELL YOU HOW TO FEEL

We normally think of the brain and neurons of the central nervous system as the signaling pathway that produces our emotions. Although the autonomic nervous system is separate from the conscious actions of the brain, the two systems exchange signals continuously. New studies even show that there is a mini brain within the heart, *the intrinsic cardiac ganglionated plexus*, where sensory neurons from the heart surface meet the autonomic neurons, and which has some processing and memory ability.[1] First and most directly, these interactions are how the effect of emotions is felt on the heart, especially fear or anxiety. We react to the sight of a spider or an irritating coworker with the same response as to the saber-toothed tiger of our evolutionary past. I remember lying on a bed wearing a heart rate monitor, waiting for some minor but uncomfortable surgery, thinking I was completely relaxed but watching my heart rate soar as the doctor entered the room.

But newer research has shown that the heart in turn controls our emotional responses. Sensory neurons send information back to the brain: fast and erratic heart beats can amplify (or even trigger) panic attacks. Even playing back to people a recording that they are told is their heartbeat, but is much faster, can bring on a panic attack. However, examples have been reported where preventing heart rate increases can reduce or even abolish feelings of fear! This has been seen in people taking beta-blockers, or with a failure of the autonomic nervous system.

Clever experiments have been able to detect changes in emotional state within one normal resting heartbeat.[2] When the heart is fully contracted (or in systole) the blood pressure is at its highest and the baroreceptors are firing off electrical signals at a fast rate. When the heart relaxes (diastole), the blood pressure is at its lowest and the baroreceptors are quiet. You usually know this cycle from your blood pressure measurements—the high of around 120 and the low of 80 for every heartbeat, if you are in good health. If we flash up images of frightened faces to research subjects under experimental conditions this sets off emotional alarm signals: we are social animals and respond quickly to fear responses in others. This is how panic is spread and amplified in crowds. Researchers linked the timing of flashing the frightened face images to the highs and lows of a single heartbeat. When the heart was in systole and baroreceptor firing high, the fear responses of the research subjects to the terrified faces was amplified. In diastole, the response was dampened. So, the heart is showing us what to feel: your body has sensed a danger, but it is the autonomic reaction that has amplified the emotion. You can see how this bypassing of the conscious cognitive processes might be an advantage, giving you that extra boost to flee the threat.

Heart rate responses to images of faces that registered disgust yielded a similar (but less dramatic) pattern to that provoked by the fearful faces. However, flashing images of faces with happy or neutral expressions produced no effect. Researchers were surprised to find that the heart rate response was completely different for images of faces in pain: emotion seemed to be dampened when baroreceptors were firing fast, not amplified. This makes sense when you think that the primary drive in the face of a threat is to get away. Pain can be ignored in favor of survival. This is the case not only for life-threatening events: anyone who has played a competitive sport can tell you that they don't feel the pain of an injury until the game is over.

RUNNING FOR YOUR LIFE—THE STARTING GUN!

Because of this reflex aspect of the response to fearful stimuli, the effect of sympathetic stimulation on the heart is extremely fast. Next time you are watching a horror film, take your pulse regularly. Then after a sudden frightening moment take it again immediately—you will see an almost instantaneous effect. Or if someone opens a phone conversation by saying "I have some bad news for you," I can guarantee your heart rate will be soaring by the word "for." This is an amazing achievement when you realize the chain of events that must happen before the full effect of sympathetic nervous system stimulation is achieved.

First, adrenaline is released from a nerve terminal onto the surface of the cardiomyocyte, since this is the muscle cell that will do all the work. Sympathetic nerves or neurons touch the surface of each cardiomyocyte at multiple points called synapses: one neuron will crawl over many cells and one cell will receive and integrate signals from up to four neurons. As with the brain, neuronal contacts continually form and reform. Neurons release neurotransmitters, which are small chemical messengers that leap across the gap to the cardiomyocyte. Although the movement of neurotransmitters across the synaptic gap is slower than the electrical impulse traveling down the neurons, it is still incredibly rapid. For the sympathetic nervous system, the transmitter noradrenaline acts on the beta-receptors; for the parasympathetic nervous system, acetylcholine acts on the separate muscarinic receptors. Muscarinic receptors are named for the deadly mushroom Amanita muscaria, which produced muscarine, the first parasympathetic compound ever discovered.

The sympathetic system also has another source of neurotransmitters, a modified nerve terminal above the kidney that over time has become a strange pea-shaped structure called the adrenal gland (the source of the name adrenaline). This releases adrenaline and noradrenaline into the blood stream to reach all organs rapidly. It also releases hormones involved in longer-term stress control, such as cortisol and corticosterone. When the sympathetic neurotransmitter messenger is released, it entwines with receptors on the surface of the cardiomyocytes. Receptors are proteins that match the shape and electrical charge of the neurotransmitters, allowing the two to bind together tightly. There are separate receptors to fit the many thousands of hormones and neurotransmitters through the body. This creates an intricate and specific mechanism to control individual cells and organs in different ways.

Receptors for adrenaline and noradrenaline (the beta-receptors) control the kickstart to the heart. The large protein receptor wraps itself around the small molecule neurotransmitter, changing its own shape at the same time. This transmits a message into the cell to set it in action. Each of these messages boosts the activity of many enzyme proteins, which in turn activate many more to amplify the effect. The proteins control the pacemaker and the force-generating mechanisms, pushing the heart to work faster and harder. It is amazing to think that all this can happen between the words "bad news" and the word "for"!

WHAT MAKES YOU STRONGER CAN KILL YOU AS WELL

Adrenaline helps you run fast; this is good for getting out of danger. Running is also good because exercise makes your heart stronger. Both these statements are true, but the sympathetic nervous system has a darker side. One of the most heart-breaking tragedies is the sudden death of a young child. We can describe a typical case, let's call the child Sam. Sam was a wiry and athletic boy, always active and playing football with friends. The coach could see he had a real talent and got him onto the junior team of the local football club. His talent developed as he got older: at 16 he had grown 8 inches in a year and was outpacing all his opponents on long, strong legs. He took to the training enthusiastically, putting in lots of time on running and strengthening exercises. Then during a match, without warning, he collapsed suddenly on the pitch and could not be revived. There had been no warning signs.

This is a horribly familiar situation and is due to a severe disturbance in the heart rhythm, ventricular fibrillation, that is so bad that blood stops pumping from the heart. It causes cardiac arrest and if not swiftly treated results in sudden cardiac death. The ventricle no longer contracts in a coordinated way—different areas are now contracting and relaxing at different times and speeds—and so it wriggles and churns without ejecting blood. Surgeons who may see this during an operation when the chest is open describe it as the "bag of worms" appearance. It can happen during a heart attack but, as with the young athletes, can be triggered by adrenaline separately and without any other stimulus. It can happen even in sleep. Professor Mary Sheppard, the expert cardiac pathologist who founded the charity Cardiac Risk in the Young knows too many families torn apart by this terrible experience. She has said, "There's nothing more devastating that can happen than waking up one morning to find your child dead in bed. It's appalling for families."

What's going on with adrenaline here? Why has something lifesaving suddenly turned on us and sent the heart's rhythm into a chaotic dance? There are two sides to these dangerous disturbances in rhythm—which we call arrhythmia—and they are caused by the immediate and the long-term effects of sympathetic stimulation. Within moments, adrenaline can trigger individual abnormal beats by overstimulating the cardiomyocytes and overcoming their natural rhythm. In a normal heartbeat there is an unresponsive time for the cardiomyocyte just after a beat, where it can't be stimulated again. This is called the refractory period. Adrenaline can make this refractory period finish early and so the next beat happens too soon—it is called an ectopic beat.

In fact, you may have experienced an ectopic beat and not known what it was—they are quite common. Because the first two beats are close together there is a long gap before the next one and your heart fills with more blood than usual. There is a mechanism that matches the force of beating to the stretching of the heart muscle when inflated by blood. Because of this, the delayed following beat is extra strong and feels like a thump in your chest. It may even make you cough. Don't worry, this beat by itself is not dangerous, though should be investigated if it happens very often or is troublesome. More problematic are periods of rapid beating or trains of irregular beats, resulting from an overload of the cardiomyocytes with calcium. Adrenaline can directly cause this calcium overload, as it tries to push the heart to its maximum.

The fight-or-flight response is an emergency reaction and did not evolve to be turned on all the time. Adrenaline (and noradrenaline) can also cause actual damage to the heart muscle when they are present for too long, or when their concentration becomes too high. If the calcium overload of the cardiomyocyte is too extreme or too prolonged, it triggers death mechanisms in the cell. Long-term effects of adrenaline are large-scale or patchy alterations in the heart muscle producing turbulent patterns of electrical flow.

In the laboratory, when we give a high adrenaline burst to a cardiomyocyte lying in a dish it stops beating rhythmically and instead waves of contraction sweep across the cell. Then these intensify so that the cell is wriggling like a landed fish. Finally, the cell shortens drastically as the muscle fibers "hypercontract" and buckle over each other and within seconds can reduce the cell's length to a quarter of what it was. The cardiomyocyte membrane bubbles and bursts with the strain, with large blisters pushing out from the surface, and the cell contents are spilled out. When this happens within the heart the cardiomyocyte is lost forever: it becomes replaced by scar. Patchy scarring changes the path for electrical flow across the heart from a smooth circuit to a maze of tiny paths, carved around the islands of damaged muscle where the current moves slowly. The electrical impulse winds through erratically, meeting itself and amplifying or canceling out. Sometimes "re-entrant" circuits arise, where a single stimulus

circles around a damaged area indefinitely. Ectopic or irregular beats are particularly lethal in this wasteland of damage and
destruction, causing irreversible ventricular fibrillation.

Boys like Sam, who collapse on the football field, often
have a condition where the heart wall thickens, and this can
also cause re-entrant circuits to start suddenly when the heart
is stimulated. Tragically, there is no outward sign that would
have alerted his parents or doctors that he was at a special risk.
Not only physical exercise can trigger this, but intense emotional states as well. A broken heart is a real phenomenon.

DYING FROM A BROKEN HEART

When the heart goes into ventricular fibrillation the victim
usually loses consciousness without pain. This is a cardiac
arrest and is fatal without immediate use of a defibrillator.
Sudden cardiac death caused by these arrhythmias can happen
separately from a heart attack. It is not only the young athlete
with a mutation who is in danger. In fact, the syndrome is
relatively rare in children, but much more common in middle-
aged adults where it accounts for around half of all heart disease deaths. Unaccustomed or extreme exercise in later life can
turn out to be deadly—the middle-aged husband shoveling
snow; the mid-life crisis executive suddenly taking up squash;
even unexpected cases—James F. Fixx, author of *The Complete
Book of Running*, died of a heart attack at age 52 while jogging.
Also the "fight" element of the fight-or-flight response can
be deadly, as anger and arguments are a very strong trigger.
Immobilization is a powerful stimulus and deaths while under
arrest or while in police custody may be closely related to this
reflex adrenaline release.

In fact, the evidence points more toward strong and sudden
emotion in general as a possible trigger. The most striking fact

is that you are statistically more likely to die in the period soon after your spouse or loved one passes away. The period of risk peaks immediately after the event and the increased risk can be measured for about six months. Debbie Reynolds died the day after her daughter Carrie Fisher, and Johnny Cash died four months after his beloved wife June. The body reacts to this intense emotion the same as to a physical threat to life. This has been termed "broken heart syndrome." The release of adrenaline is the same, the action on the heart is the same, and the outcome of arrhythmia and sudden cardiac death can be identical. What we know now is that reactions to strong emotion are intimately bound up with very primitive responses to danger. They trigger instant activation of the sympathetic nervous system, release of adrenaline and noradrenaline, and stimulation of the heart, setting in train the events that both disturb the rhythm and produce cardiac damage. As we will see, there are two different diseases called broken heart syndrome, with very different outcomes biased by whether you are male or female. We will return to broken heart syndrome in the next chapter, but it's important to explain now why this is part of our physiology—and what your personal risk is likely to be.

THE SPECIES VS. THE INDIVIDUAL

Why do we have this time bomb inside us—a system that can kill us in multiple ways, without warning or by slowly destroying the heart? We have to look back to the times when we evolved, when our ancestors died often from childbirth, accident, and attack. This was long before our worst enemies became the sofa and the attractions of calories, alcohol, and tobacco.

Think of a mouse that has been genetically altered to remove all the beta-receptors: it cannot respond to the adrenaline rush of danger. It lives a perfectly long and happy life in the colonies in university animal housing facilities. But the clue to this paradox is that this mouse would not last very long in the wild world. Imagine a cat watching a nest of mice. They sense it and scatter—the cat pounces forward to chase them. The mouse with the adrenaline system intact will double its heart rate and force, divert the blood to its muscles, and shoot away into the distance. The mouse without beta-receptors will be snapped up straight away and bitten in half. All the babies it might have had will be lost. Such mice are much less likely to survive to adulthood and reproduce, so the genetic makeup (or genotype) that lacks beta-receptors will be lost to future generations. Although there is a small chance that the normal mouse with the full beta-receptor response to adrenaline might also die of arrhythmia caused by the fright, there is a much larger chance that it will survive the attack and go on to have a big family. So, the species overall benefits from more mice, at the risk of losing one or two to adrenaline overload: mousekind is saved!

In our evolutionary past we had the same dangers and perils to overcome, and the extreme flexibility of the autonomic nervous system would have given a selective advantage for humankind not only for fight or flight, but also for serious injury. Extreme blood loss after wounds produced by attacks or falls causes a drop in blood pressure. This is sensed by the body as a mismatch between cardiac output and the need for blood flow to organs. Here, adrenaline closes off blood vessels to prevent further loss; instructs the kidneys to save water to increase blood volume; and stimulates the heart to beat harder to make up for the reduced blood. For these emergency rescue situations, as well as for many other aspects of our physiology, evolutionary mechanisms can be thought as a trade-off between the good of the individual and the survival of the many.

SHOULD I TAKE DRUGS TO BLOCK ADRENALINE— JUST IN CASE?

To have an active sympathetic and parasympathetic nervous system is a sign of cardiac health, but in certain circumstances adrenaline can be deadly. How can we balance this risk? In most parts of the developed world we are more like the laboratory mice, living an orderly, sedentary life and no longer threatened by predators. Do we need adrenaline? Should we give everyone beta-blockers, drugs that prevent the action of adrenaline at the beta-receptor?

The answer is not straightforward. For a typical young, healthy person the risk of dangerous arrhythmia is not high. The heart beats 100,000 times a day and can remain steady for 100 years or more—as it must when even four minutes without beating will kill you. We have seen how the heart has evolved exquisitely robust mechanisms to prevent this happening. Its structure as a functional syncytium for example—the connection of each cardiomyocyte to many others—damps down arrhythmias originating in a single cell, and it needs a larger number of cardiomyocytes to be disrupted before the whole heart is affected.

One thing that is clear is that exercise, especially vigorous exercise, is good for the heart in the long term.[3] The evidence for this is increasing every day. In large groups of disease-free people, a difference in activity equivalent to a 30-minute brisk walk three days a week leads to a 7 percent drop in death rates within six years. In patients with prior history of heart disease, the difference is double that. In fact, mortality in the sedentary disease-free group was worse than in the exercising group with disease. We need adrenaline to exercise properly, and the health of the heart is directly related to this effect. So, the risk/benefit analysis for the disease-free population is clear: keep the adrenaline.

However, this is not the same for people with advanced heart failure, where adrenaline is a strong extra risk. Heart failure is a disease with death rates equivalent to some of the worst cancers, and half the people with heart failure will die from arrhythmia. Adrenaline creates extra danger because it not only triggers the arrhythmia but contributes to further damage. The body responds to the weak cardiac output as if it were an emergency, such as hemorrhage, and releases adrenaline continually and in large quantities. What was a short-term emergency measure is now a long-term burden. Many cardiomyocytes are overloaded with calcium and further cell death makes the damage and dysfunction worse. So, the risk/benefit analysis for the part of the population with existing heart disease is clear: block the adrenaline. Therefore beta-blockers have become a mainstay of heart failure treatment.

Our real uncertainty about long-term beta-blocker treatment lies when the heart might have underlying faults, such as with undiscovered genetic mutations, but not yet show overt heart disease. There is a large group of mutations where beta-receptor activation will combine with the main gene defect to produce arrhythmia—this is the case for some of the sudden infant death syndromes, for example. One of the most common sets of mutations produces hypertrophic cardiomyopathy (thickening of the heart wall), which happens in one of 500 births. Going back to our young athlete, Sam, who collapsed on the playing field, hypertrophic cardiomyopathy is a common underlying cause of heart thickening. Because his heart wall was normal until puberty, the condition had not been detected. Only after the incident was his enlarged heart found. After some high-profile incidences of fatal or resuscitated sudden cardiac events, many teams are starting to have their players screened for hypertrophic cardiomyopathy.

Should we give everyone beta-blockers just in case? The answer is no—prescribing beta-blockers to the whole population to prevent a rare event is ethically and practically unacceptable. As well as their side effects such as insomnia and fatigue, they stop your heart speeding up. This can be a barrier to exercise, which we know is one of the best things for your health. Adrenaline is involved in emotional responses to excitement and so it's not surprising that people often report feeling "flat" on the beta-blockers. This risk-to-benefit ratio for rare but serious effects is an exceptionally difficult dilemma for public health strategy. Looking to the future, we can only hope that large-scale predictive genetic screening can help us target more accurately who might benefit from anti-arrhythmic protection.

So, we have reviewed in this chapter the damaging effect of adrenaline, but also how the benefits of the sympathetic nervous system response have been essential for our evolution and are still integral to the vitality of our day-to-day lives. In chapter 7 we will explore a different side to adrenaline through another disease, also called broken heart syndrome, but from which most people don't die. This is such a strange syndrome that it wasn't even recognized as an illness until very recently, and it can still be written off by clinicians as patients' imagination. But it holds a critical key to our understanding of how the body protects itself from harm.

CAN YOU (NOT) DIE OF A BROKEN HEART?

The first report of a strange new heart disease began in 1990, among the chaos and carnage of a huge earthquake in Japan.[1] Hospitals were inundated, not only by the injured from the earthquake damage, but also a wave of people with suspected heart attacks. This is a phenomenon that had been seen before around the time of major disasters and yet the cause had remained a mystery. But the difference from other such disasters is that it was happening in Japan, a country with a significant number of high-tech hospitals. In Hiroshima City Hospital, doctors were now using state-of-the-art imaging techniques such as coronary angiography to visualize the coronary arteries—the blood vessels that supply blood to the heart muscle. Cardiologists inject a contrast agent to see whether there is a blockage from a clot or ruptured plaque that might be obstructing the vessels and causing the heart attack. But in a new procedure, they also moved the injection of contrast to the ventricle to show the shape of the heart as it contracts. In one group of patients they saw two things that amazed and puzzled them.

First, there was no blockage, despite the patient having all the pain, ECG changes, and blood markers that point to a heart attack. Second, the heart as it contracted showed a shape they had never seen before. The top of the heart near the atria (called the base) was contracting extremely vigorously, so

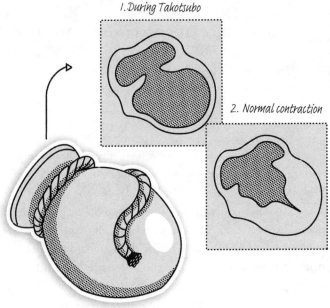

Figure 7.1
The normal heart is shown at the peak of contraction, where the left
ventricle walls contract to expel blood. At the same point, the apex of
the heart in Takotsubo syndrome is thin and not contracting. It has
been compared to the shape of the traditional takotsubo octopus pot.

much so that it was even blocking the outlet of the left ventricle. But the bottom of the heart (or apex) was almost immobile. This produced a shape like a balloon on the X-ray (figure
7.1) or like a narrow-necked pot. It reminded the Japanese
doctors of the pots, or *takotsubo*, used for trapping octopus.
The narrow neck imprisons the creature after it crawls in. This
has given the disease the eye-catching name of Takotusbo syndrome, also sometimes called stress cardiomyopathy, and it is
the second broken heart syndrome.

Two other features made this syndrome stand out to the clinicians treating these patients. The first was that their patient group consisted of mainly women. That's quite unusual for heart disease in general, which is more expected in men—at least for younger or middle-aged people. In all age groups the men outnumber the women, by more than two to one in ages below 64 years. It is true that the gap narrows in later life, after menopause, and in fact many of the women with Takotsubo syndrome were postmenopausal. When we add up the figures over many studies, an astonishing 80–90 percent of patients in the Takotsubo group are women.[2] This is a huge majority, and one that had never been seen for a type of heart disease before. An even more striking feature is that many of them recovered completely, with no sign of cardiac damage. They went from an emergency admission with severe chest pain and acute heart failure, to walking out of the hospital disease-free, sometimes in a matter of days. Most never had a problem again. To scientists and doctors this is highly intriguing. We know people with this severity of acute heart failure would usually have a very poor outlook. If the cardiologists had not seen the strange shape of the heart and its bizarre contraction on the X-ray, then we would have thought that the patients had been imagining or misinterpreting the symptoms.

It's very sobering, in fact, to think that women suffering from Takotsubo syndrome might have been dismissed as malingerers before such imaging was possible. You can see how a middle-aged woman, distraught after an upsetting event, coming into hospital with chest pain and collapse but recovering quickly, might have been seen as having a "hysterical" illness. Taking this even further, many people who have panic attacks say, "I thought I was having a heart attack, but I was fine after a while." Maybe they were Takotsubo sufferers too?

A MISSED PENALTY KICK

One remarkable incident really highlights the different responses of men and women to extreme stress.[3] A family of ardent Chilean football supporters—a husband and wife and their three children—were watching together the dramatic battle between Chile and Brazil for a place in the next round of the FIFA 2014 World Championships. There was a tense penalty shootout. The last and deciding kick hit the goalpost, ending Chile's hopes. The family immediately erupted into a ferocious quarrel that raged until the 58-year-old father clutched his chest in severe pain. He had a cardiac arrest and was rushed to the emergency room, where he was found to be in ventricular fibrillation and had two rounds of defibrillation. Sadly, however, the event finally proved fatal.

Just over an hour after he had the cardiac arrest, his 64-year-old wife began to feel chest pains. She showed the same ECG changes as her husband but did not go into ventricular fibrillation. Within an hour of her husband, she was being treated in the very same angiography laboratory, where they were able to see general furring of her arteries, but no sign of a block. Instead, there was the pattern of a ventricle contracting strongly at one area but almost immobile in another: this was clearly a case of Takotsubo syndrome. This time it was not the apex/base difference seen, but a ring around the middle of the heart that was contracting very weakly. She was given supportive treatment only and her heart gradually became normal over time in the hospital. She was discharged after a full recovery.

The stressors here are important. Clearly the football match was the first trigger, and the main one for the husband, with possible additional triggering from the argument. Watching football, and sports in general, has often been associated with cardiac events. One study reported that mortality due to myocardial infarction and stroke increased when a local team played at home.[4] The penalty shootout is a particular stress—both winning and losing! Myocardial infarction increased by 25 percent on June 30, 1998, which was the day that England lost to Argentina in a penalty shootout, and on the following two days.[5] When the World Cup was held in Germany, on days of matches involving the German team the incidence of cardiac emergencies was 3.3 times the normal rate for men, with an overall increase of arrhythmia also threefold.[6] However, it was the *tension of the game* rather than the result, since the effect was the same whether the team won or lost.

Interestingly, the same German study showed that for women the increase in cardiac events during a German game was not so marked, with a rise less than twofold. For the wife in the Chilean family then, it is possible that the match was not the only cause. The stress of seeing her husband with a cardiac arrest was most likely the strongest trigger, coming so soon after the game and the fierce argument. In her husband the extreme stress of both the match and the argument had caused fatal arrhythmia. Similar emotion in his wife did not set in motion that same burst of arrhythmia. In fact, the excessive heart contraction she experienced was damped down by the Takotsubo mechanisms.

TAKOTSUBO—A NEW BROKEN HEART SYNDROME

At first, Takotsubo was thought to be a Japanese disease, and a very rare event, but gradually the consensus has changed. There were only two published scientific papers recorded in 2000—this leapt to nearly 300 in 2010 and they came from all corners of the world. Cardiologists now have much greater access to advanced imaging, and ventricular shape assessment is routine. Takotsubo is known as a syndrome, which describes a condition characterized by a set of associated symptoms. Like many syndromes, it was first identified by its most extreme manifestation: catastrophic stress; ballooning of the heart apex; acute heart failure; no sign of coronary blockages; complete and rapid recovery. With time, syndromes often expand their descriptions: now we know it may be chronic rather than acute stress; the heart may contract oddly in other ways; there may be some coronary atherosclerosis; there is an early mortality of up to 5 percent and the recovery may leave patients with a residual heart impairment. It also may be more widespread than was first believed because of the different ways that it can present when patients are admitted to the hospital.

Many of the stressful incidents strongly paralleled those triggering sudden cardiac death (SCD): disasters like earthquakes and tsunamis almost always see a rise in both Takotsubo syndrome and cardiac arrest. As with SCD, extreme physical exertion is a danger point.

Arguments are also frequently reported as the cause that sets off a Takotsubo event. Bereavement again is a clear trigger, especially the loss of a spouse or a child. Takotsubo is also associated with the "anniversary reaction" where a heart attack happens around the anniversary date of a past traumatic event. It is this linkage with extreme grief plus the highly distinctive distortion of the heart on the images that labeled this new phenomenon as "broken heart syndrome" or stress cardiomyopathy.

But the classic Takotsubo syndrome patient is a woman in her 60s. Let's call her Mary. She is active and sociable with a wide circle of friends who lean on her for advice. She has a close but complicated family: maybe her grown children are divorced, and she is helping with the grandchildren. More often now she is shuttling back and forth to look after elderly parents while perhaps juggling a job, too. Her days are busy and draining. She certainly doesn't think much about her own health, dismissing aches and pains as the inevitable part of oncoming age. Her husband is her rock and comfort in all her troubles, and they have supported each over the problems of the years.

Then her world falls apart. Her husband dies: maybe suddenly, with no time to say goodbye, or maybe after she has nursed him through a long and difficult illness. The funeral arrangements go past in a blur; she is in a kind of numb shock. Her thoughts are for her children and grandchildren, to comfort them in their distress.

During the funeral or the following days she feels increasingly unwell. At first, she doesn't notice how bad she feels against the background of her grief and sorrow, but as the rush dies down the exhaustion and nagging chest pain get worse. She soldiers on, not wanting to cause her family any trouble, hoping that the feelings will subside. She may collapse, or finally admit how awful she feels, and is taken by her worried family to the hospital. Thankfully, this facility has doctors who take her symptoms seriously and act rapidly.

Doctors know that "time is muscle," and saving heart muscle has a profound effect on the future lifespan and quality of living for the patient. Highly sophisticated imaging is now common. Fewer people are dying, and many get back to a good quality of life. Mary is quickly taken to the ER and the doctors see that her ECG and symptoms ring alarms bells that point to a possible heart attack. They take her to the catheter laboratory to give her an angiogram, where the blood vessels are visualized by a live X-ray during which a contrast agent is injected into the coronary arteries. What they are looking for is the culprit blood vessel that has been blocked by a blood clot or an atherosclerotic plaque (a thickening of the vessel wall containing cholesterol) that has suddenly burst. The cardiologist does not see any blockage, but there is not a great deal of surprise. This does happen—maybe the clot has broken up or the blockage was due to the blood vessel having a spasm. However, the bizarre motion of the heart alerts them to something very different happening, and the weak contraction of the apex is a real concern.

What happens next actually depends on whether the doctors have heard of Takotsubo syndrome. Knowledge is spreading, but while hospitals in major cities are expert in a wide variety of heart conditions, including this one, smaller local hospitals may have seen only a few Takotsubo cases. In this hospital they are unsure what to do. Mary is put under observation while they decide and give supportive care that keeps her condition stable. Now that the emergency has subsided, they can take some time. Mary begins to feel better over the next few days, so she has cautiously begun walking about the ward. A week or so later she is examined again, and her heart is functioning better, although not completely normally. The pattern of apex/base difference is not so evident. Mary has been lucky: a small percentage of Takotsubo victims die of complications during the acute first illness, but like most others she has passed through the immediate danger and the prognosis is good.

Another week passes, and the improvement continues. Mary is longing to get home and protests that she feels "fine"; meanwhile new patients are lining up for her bed. She is discharged and goes back to her busy life with support and care from her family. Months later she comes for another check-up. Her heart is functioning adequately—not like a 20-year-old but not so unusual for her age. The doctors chalk this one up to experience, just happy that the outcome was good. Of course, no one had examined her before the attack, so there is no way of knowing whether she has truly recovered completely. She feels okay, with less energy and enthusiasm for life than before but, well, that's to be expected after losing her husband. Has she really recovered completely? It's difficult to tell.

SEX, DRUGS, AND ROCK 'N' ROLL

Mary's Takotusbo experience is the classic case, and it's a story that clinicians hear again and again. However, the tentacles of this octopus syndrome seem to spread further as the reports come in from all parts of the world. At first, the papers being published on Takotsubo syndrome described these kinds of triggering incidents linked to disaster and grief. Often, the triggers were very like those for SCD, such as bereavement and trauma.

Then other anecdotal reports began to appear where the triggers were a bit more unexpected. An 80-year-old man having an illicit affair with a much younger woman—maybe a mixture of emotion and exertion? A 69-year-old woman after a two-hour dancing session—this could again be exertion with an added factor of emotion.[7] But neither of these two cases seem so linked to sadness and heartbreak. Then came a study that even happy events could cause Takotsubo: surprise birthday parties, for example, or the wedding of a son.[8] Of course, you could argue that these occasions elicit strong and very likely mixed emotions. And why do these emotional and physical triggers cause Takotsubo syndrome and not sudden cardiac death?

Then came reports of drug-induced Takotsubo syndrome. Various kinds of energy drinks seemed to be triggering it, especially those that contained caffeine and taurine. "Male enhancer" pills with pre-Viagra drugs have been implicated in some cases.[9] There were several reports that a dental injection of local anesthetic had been involved.[10] Even in hospital conditions, some of the tests and therapies had unexpectedly induced a Takotsubo-like syndrome. One of these is the stress test for cardiac function when a stimulant called dobutamine is infused to see how well the heart can respond, and there are multiple reports of this as a trigger.[11] Takotsubo has also been set off by induction of general anesthetic during routine operations.[12]

Asthma drugs seemed to be particularly problematic—in one study 44 percent of the Takotsubo patients were found to be asthmatic.[13] Antidepressants have been found to be a factor both during their administration and during withdrawal.[14] A single injection from an EpiPen, used for anaphylactic shock, has been seen to precipitate cases of Takotsubo syndrome in recipients.[15] Hospitals also saw the syndrome associated with other diseases and treatments—this is called secondary Takotsubo. Disease or drugs that raise thyroid levels to a toxic level and dangerous septic shock have caused cases.[16] The link to brain function is seen through Takotsubo caused by head injury and bleeding into the skull.[17]

Gradually, the picture is starting to clear. Scientists and clinicians have combed through this wealth of information to discover clues to the mechanism of Takotsubo syndrome. The evidence for adrenaline, or the class of related compounds (called catecholamines) in general, as the culprit began to rise: the EpiPen is epinephrine/adrenaline; dental injections of anaesthetic also contain adrenaline. Head injury and brain hemorrhage also are linked to increased activity of the sympathetic nervous system. Thyrotoxicosis increases heart rate because the number of beta-receptors goes up, and so the responses to catecholamines are much more sensitive. Dobutamine stimulates the beta-receptors in the same way as noradrenaline to test out the heart's ability to beat harder. Caffeine acts inside the cell to stop the breakdown of the messenger compounds that produce the effects of beta-receptor stimulation, and so enhances and prolongs the action of adrenaline. *The smoking gun is the beta-receptor.*

But why are these patients recovering? Why Takotsubo syndrome and not SCD? One singular piece of evidence that opened our eyes came from the asthmatic responses. In severe asthmatic attacks the sufferers can be given high doses of adrenaline, but for everyday relief they use drugs such as salbutamol or salmeterol that only stimulate one subclass of the beta-receptor called the beta2.[18] Is there something special about the beta2-receptor that links it to Takotsubo?

OF HOT MICE AND MEN

It's surprising the role that serendipity plays in science. The impression that scientists give is that everything is logical and planned, but often many factors converge to bring us to an understanding of disease. Our first clue to the link between the beta2-receptor and Takotsubo came when we were investigating mice that had been given extra copies of the beta2-receptor by genetic engineering. It was a very hot summer that year in the United States, where the mice were bred, and we had trouble shipping them over to the UK. Several times they were brought out onto the tarmac to get onto the plane, but it was too hot to take off. When they arrived at our laboratory, the effects of the added beta2-receptor seemed to have disappeared. Their hearts did not respond to adrenaline with an extra-strong increase in rate, as the US scientists had seen, but now adrenaline even reduced the rate. But also, the mice could tolerate much more adrenaline without getting arrhythmia.[19] It was as if the high adrenaline was now slowing and protecting the heart—completely opposite to its normal action! We kept checking our results thinking we had mixed up the cages or made some other terrible error.

Luckily, a second group in the UK headed by Sir James Black (who had won the Nobel Prize for the invention of beta-blockers) had also received these mice from the United States and saw the same thing, so I knew it wasn't our mistake.[20] Years of investigation followed, where we went back and forth between our laboratory models and the human cardiomyocytes from our patients and donors. We used advanced imaging methods and molecular tools to tease apart the exact receptor mechanisms that were behind this unexpected switch in function of the beta2-receptor. Finally, we understood that the beta2-receptor was now linked to a completely different set of intracellular signaling molecules. It was giving different instructions to the cardiomyocyte, suppressing the contraction but also activating cardioprotective mechanisms that prevented cell death and arrhythmia.[21] The very strong stimulus of the extra beta2-receptor and stress and heat of the trip had activated a mechanism in the mice that switches the beta2-receptor from its usual cardiostimulant (but eventually damaging) to a cardiodepressant (but protective) action.

We then tested out the hypothesis on normal rats under anesthetic and we were able to show that a very high concentration of adrenaline could cause the same kind of switch.[22] A dose of adrenaline, equivalent to an EpiPen in humans, first had its usual action to increase heart force, then after about 15 minutes changed to a depressant phase that lasted for about 45 minutes (until the adrenaline finally washed away). Because the beta2-receptors are more concentrated in the apex of the heart, the depression was more obvious there. On the monitor, this looked exactly like a Takotsubo patient's heart. With one dose of adrenaline, we had reproduced the clinical syndrome of Takotsubo cardiomyopathy!

TAKOTSUBO AND SUDDEN CARDIAC DEATH— TWO SIDES OF THE SAME COIN

With our anesthetized rat, we had a model where we could try to understand the disease and look for a cure. From our experiments, we now knew how to block the new adrenaline signaling that we thought led to the localized depression of heart function. We tried this out in the model and it worked—we could prevent the apical ballooning. But suddenly, the rats started to die under the anesthetic. We were dismayed to find that the depressed heart function had been replaced by deadly arrhythmias! By preventing the Takotsubo syndrome we had triggered sudden cardiac death. This was a big setback for treating Takotsubo—in fact it showed us that this was something we should avoid at all costs. But it did confirm the idea we had formed from our work with the US genetically altered mice: that adrenaline was protecting the heart from further damage in this new signaling mode. So, for one adrenaline signaling pathway you get arrhythmia and sudden death, but in the other, this is converted to Takotsubo syndrome and an acute but reversible heart failure. After an extreme stress event, you might think of Takotsubo syndrome as the *least worst outcome*—a kind of aborted sudden death episode.

But why are postmenopausal women the main targets for this syndrome? While 80–90 percent of SCD patients are male, 80–90 percent of Takotsubo patients are female.[23] In our rats, like humans, the male rats are much more likely to have arrhythmia with the high adrenaline, but they can also get Takotsubo. Young female rats have very few arrhythmias with the adrenaline dose we give. What about older female rats? Most mammals don't go through menopause—curiously, only humans and some species of whale do. To test the role of estrogen we had to reduce or block estrogen levels in the rats by using drugs. This made the females respond much more like males to the adrenaline dose. Estrogen is clearly a key player.

An intriguing link to menopause is the ability of adrenaline to raise body temperature and produce something like a hot flash. It is also known that Takotsubo syndrome is more likely to occur in summer while most heart disease spikes in winter. The phenomenon of increased body temperature after a surge of adrenaline release is well known in older studies and has been called "emotional fever." We had seen the signaling switch in our mice exposed to high temperature during travel. In another example of serendipity, we found that moving our rats off their warmed beds during the anesthetic (to stop this rise in body temperature) prevented the Takotsubo effect after the adrenaline injection. Is temperature also the link that produces Takotsubo syndrome in the extreme fever of sepsis?

Why are males more susceptible to adrenaline-induced arrhythmia? Maybe a better question is why do young females need to be protected more strongly than males from adrenaline? My hunch is that the extreme physical and emotional stress of labor and birth will overwhelm the mother with adrenaline stimulus, and that extra protection is needed. From a biological point of view, protection of the mother at the moment of birth represents a very strong evolutionary pressure. But frankly, it is hard to design ethical experiments to test this.

WHAT CAN WE LEARN FROM TAKOTSUBO SYNDROME?

So, now we know that there is a way that adrenaline can be made safe in high doses. In fact, not only safe, but also protective of the heart. The signaling switch that does this is present in both men and women but is much more active in women. Maybe it is protecting all of us and, if we didn't have it, sudden cardiac death would be much more common. Is there a way that we can boost this effect or mimic it with a drug?

If we take our findings to their logical conclusion, then estrogen would be the ideal agent to protect against sudden cardiac death. For postmenopausal women, the anecdotal evidence is that Takotsubo is seen less in those on hormone replacement therapy. However, the general protective effect of being female for heart disease has been known for some time, and somehow the idea of estrogen supplements for men has not really caught on! Surprisingly though, one of the most common drugs for heart disease and arrhythmia may have already harnessed some of the protective pathways in Takotsubo syndrome. These are beta-blockers.

We first had the breakthrough idea about this when one of the postdoctoral researchers in my lab did an experiment to put a high dose of beta-blockers directly on a human cardiomyocyte from a failing heart. (I distinctly remember forbidding him to do this experiment because there was clearly no point, nothing would happen—which shows you how nature can always surprise.) What in fact happened was that the beta-blockers made the beating force of the cardiomyocyte dramatically decrease, like the cardiodepressant effect of Takotsubo. This was theoretically impossible, because there were no sympathetic nerves in the dish with the cardiomyocyte and no added adrenaline. If the beta-receptors were not being stimulated, how could the blockers have anything to block? The only explanation was that the blockers were doing something extra and active to depress the heart function.

There are quite a few different kinds of beta-blockers because many drug companies have tried to outdo each other to make better ones: some of the beta-blockers had the depressant effect on the cardiomyocytes and some did not. We went back to our experiments and found that these special beta-blockers with depressant action were acting through the beta2-receptor and were using the same mechanism as seen in Takotsubo syndrome. These beta-blockers had mimicked the cardiodepressant effect of adrenaline and acted in the same way to protect the cardiomyocytes.[24]

The new action of the beta-blockers explained a puzzle we had been struggling with for a while. When the first of the beta-blockers, propranolol, was given to heart failure patients in the 1980s it proved to be lethal in some patients because it dramatically decreased the force of the heart. Because of this drug, beta-blockers were banned for heart failure for quite a few years. Clearly, propranolol has gone too far in the cardiodepressant direction and, when combined with an already failing heart, has dropped the cardiac function fatally low.

Years later, some newer beta-blockers were cautiously tried in heart failure patients at low, then gradually increasing doses.

At first, the patients would feel worse and their cardiac function would drop, but over some months they would recover and then improve.[25] Most importantly, death rates from heart failure, which would kill half of all patients within five years, were cut by 35 percent. In these drugs, the balance between blocking the arrhythmic effect of adrenaline and activating its cardiodepressant/cardioprotective pathway has been achieved. Now beta-blockers have gone from absolutely forbidden to absolutely mandatory in treating heart failure. Serendipitous design of beta-blockers has given us a drug that mimics the protective effect of adrenaline through the beta2-receptor.

WHO WILL GET TAKOTSUBO?

Takotsubo syndrome is relatively rare—of all the people who come to the ER with a suspected heart attack only about 3 percent turn out to have this syndrome. We suspect the true rate might be a bit higher, since it is not universally well diagnosed because of the variation in the syndrome, but it is still a low percentage. If you have underlying heart disease or a certain type of mutation, there can be an increased risk of SCD. But is this also the same for Takotsubo?

It is true that the risk of a second Takotsubo attack is high in sufferers who have already had one, with around 10–15 percent experiencing another one. This points to some underlying factor predisposing people to the syndrome, separate to the final adrenaline surge. Take our case of Mary: her heart function was found to be moderately impaired after the episode, but this could have been something she had been living with for a while. Perhaps this made her more likely to have an attack? There has been a search for a mutation that could link to Takotsubo, but nothing very conclusive has turned up.

We do have a clue, however, that another class of molecule, which controls how genes turn on and off, might play a role in Takotsubo. These are the *microRNAs*—the orchestral conductors of the body.

When the human genome was first sequenced in 2001 it was a surprise to everyone how few genes there were (20,000—which is fewer than a water flea has) and how much of the genome (more than 90 percent) didn't code for any genes. Because this noncoding region would not produce any of the proteins in the cell, it was known first as junk DNA and thought possibly to be left over from viruses that had infected us. But much of this is far from being junk DNA. We now know for example that the microRNAs, short sequences copied from the DNA, come from part of this noncoding DNA. MicroRNAs act to turn on and off groups of genes quickly and flexibly, to orchestrate responses to the challenges of daily life.

When Takotsubo syndrome was first identified, clinicians looked for ways to identify Takotsubo patients and distinguish them from the others having heart attacks. This was important because some of the drugs or tests for heart attacks, like dobutamine stress testing, would make the Takotsubo patients worse. Two microRNAs, it turned out, were clearly raised in the blood of Takotsubo patients but not in the heart attack victims and so were "biomarkers" for the syndrome. We gave these to our rats for several weeks and found that they would switch to the depressant response much more easily.[26] They had been primed to get Takotsubo. The two microRNAs are linked to anxiety and depression: for example, students going through the stress of exams show raised levels of these molecules. Turning back to Mary, our Takotsubo sufferer, it is possible that her complicated life and day-to-day stress had made her susceptible, so that when the adrenaline surge finally came, she was more likely to get Takotsubo.

WHAT'S NEXT FOR THE TAKOTSUBO PATIENT? FIRST DO NO HARM

By now, we have at least learned what *not* to do for the person diagnosed with Takotsubo syndrome. No adrenaline, no indirect beta-receptor stimulants, no dobutamine stress test. Possibly some stimulants unrelated to the beta-receptor might work—these help in the rat but need to be tested in people. However, the sudden and rare occurrence of Takotsubo makes a planned clinical trial very difficult to organize. Blocking the microRNAs might be a good pretreatment but, first, this kind of drug is only just coming onto the market and second, we would have to give it to many thousands of people just to protect the susceptible ones. The same thing is true for beta-blockers, which could prevent the effect of the adrenaline surge but would have to be taken by many apparently healthy people. It would be ideal if we could predict the people likely to react to high adrenaline with an extreme response, so that we could protect them with these drugs.

So, our plan is to screen as many people as possible for a Takotsubo-type pattern of cardiac function. We will start with people who have one attack, since we know they are more likely to have another. Our study will equip them with wrist monitors, like the Fitbits and Apple watches, to record heart rate variability and physical signs of stress. Patients will also record when they are stressed in a diary app. We will look for patterns of high sensitivity to adrenaline (the trigger) followed by an anomalous dip in cardiac function (the signaling switch). We will use these patterns (and with artificial intelligence methods look for others that we would not have predicted) to link them to Takotsubo recurrence. Eventually we will roll this out to volunteers to see if our idea is right that most people will show this effect to some extent, and that maybe it is part of the physical symptoms of anxiety and panic attacks. Hopefully, we can use the apps to help sufferers to understand their susceptibility and to anticipate and avoid another attack. Maybe it will also be a comfort to know that, ultimately, it is part of a natural mechanism to protect us from the risks and dangers of adrenaline.

We have seen that sex differences in a response can tell us a new truth about the biology of arrhythmia. Genetic differences between the sexes, but also the different behaviors, experiences, and life events between men and women can tell us much about heart disease in all its complex aspects. Gender fluidity, as biological sex and gender become disentangled, has fascinating insights that can make us rethink some of our assumptions. Chapter 8 will show you the unpeeling of the layers of cause and effect, as well as give you an opportunity to learn how closely your own gender and biological sex are related!

THE GENDERED HEART

SEX, HORMONES, AND SOCIAL CONDITIONING
IN HEART DISEASE

The story of the two broken heart syndromes has dramatically highlighted the different ways in which the hearts of men and women can respond. But sex differences exist through-out cardiac disease risks, development, treatment, and out-come—sometimes glaringly, sometimes more subtly. Their first origin is in the X and Y chromosomes that determine biological sex, which are not virgin territory but have been marked with the legacy of our parents and grandparents. Sex differences are maintained and magnified by the ebb and flow of hormones: in the womb before birth; during the surge of puberty and in the decline of old age. Divergence between the sexes is fine-tuned by the different stresses imposed by society, as well as the perception of gender-related behaviors. In turn, those societal expectations and interpretations are intimately related to the same biological foundations of genes and hor-mones. It is a complex web to untangle, and it has been easy to make false assumptions. We need to look to see what hard evidence exists.

First the facts. Women are less likely to suffer from coro-nary heart disease at young ages and so less likely to have heart attacks. Figure 8.1 shows the basis for this view, where young

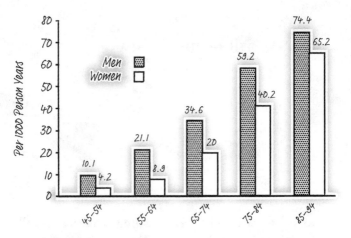

Figure 8.1
Annual number of adults having diagnosed heart attack or fatal
coronary heart disease by age and sex.

or middle-aged women have less than half the risk of a heart
attack compared to men in those age groups.[1] After age 65,
rates rise faster for women and they start to close the gap with
men. In the very elderly the balance is almost equal, although
many of the men susceptible to heart disease, who would have
been in this age group, will already have died. Women tend
to be diagnosed more slowly and have poorer treatment—
fewer stents put in to open blocked blood vessels and less of
the recommended drugs given when patients are discharged.[2]
Drugs that have been developed in laboratory studies using
male mice, then tested in clinical trials with mainly male subjects, frequently do not work in women (or need drastically
different doses).

For heart failure, the numbers of men and women are more
equal. Heart failure is the chronic disease with water retention, breathlessness, and fatigue as its main symptoms, which

develops after many kinds of damage to the heart or blood vessels and increases strongly with age. The death rate at the beginning of the 2000s was higher for men (about 30 percent), while in 2012 to 2015 it was almost the same for both sexes, mainly because of improved male survival. Women tend to be diagnosed later and survive for less time after diagnosis. The typical male pattern of heart failure is poor contraction of the heart, and this is usually related to loss of heart muscle. However, women are more likely to have a heart that contracts normally but is stiff, and therefore does not relax well between beats. This is called heart failure with preserved ejection fraction (HFpEF) and is due to changes in the small blood vessels, as well as fibrosis within the muscle of the heart. (Also, in heart attacks women are more likely to have problems with blood flow in the smaller vessels of the heart rather than a major blockage in a single large vessel.) HFpEF is often related to diabetes and obesity. Where do all these differences spring from? Is it the pattern laid down in our genes, or the lifelong differences in hormone flux? Or is it the way that gender shapes our behavior and lifestyle?

GENES AND SEX

The blueprint for our bodies lies in the coils of DNA packed into 46 bundles within the nucleus of each cell. Each bundle, or chromosome, contains many genes and each gene produces one of the proteins that make up our cells. The chromosomes gather in pairs (one from each parent) and most of these pairs are equal in size, except for one—the sex chromosome. Women have two equal-sized chromosomes, termed XX, while men inherit only one from their mother, but have a much smaller Y chromosome from their father. It is XX vs.

XY that determines biological sex. The immediate implications are obvious: men have genes in the Y chromosome that
women do not but lack the backup of a second X chromosome. The Y chromosome is very small compared to the other
45 bundles and mainly contains the program for converting
the female body template to a male one. The SRY gene on the
Y chromosome triggers the development of the fetal testis,
which releases male hormones to affect development of the
baby. The X chromosome is larger and has a wider array of
genes, including genes of the immune system.

Having two sets of genes, one from each parent, means
there must be a mechanism for the body choosing which one
to use. This is where epigenetics comes in. Genes arrive in the
fertilized egg marked differently whether they came from
the mother or the father. These epigenetic marks are chemical
groups attached to the DNA, which can control how strongly
a gene is expressed. When we were talking about diabetes, we
saw that the life experiences of the parent in terms of starvation vs. abundance of food could be passed down to the child
through epigenetic marks. This controls whether the child will
seek and metabolize food in a "thrifty" way, expecting lack of
food. (Fun fact: it's very difficult to clone a domestic cat; because
its coat colors are created by epigenetic marks, the cloned
kitten will often look very different from the original pet.)

Men only have one X chromosome and this must come
from their mother; therefore it has only the maternal imprint-
ing pattern. Therefore certain diseases (and male pattern
baldness, interestingly) are passed only from mother to son.
Daughters can escape these effects because they have a second
copy of the gene on the other X chromosome. The body randomly chooses which of two X chromosomes to use for each
gene and inactivates the other one. However, sometimes this

X-inactivation is not complete, and so women get an extra dose of that gene. When the X genes linked to immune reaction are magnified in this way it has both good and bad consequences. On the one hand, women are generally better at resisting infectious diseases. On the other hand, they are much more prone to autoimmune conditions, like lupus and multiple sclerosis, where the body's immune system is overactive and mistakenly attacks and damages the body's own organs.[3]

During the COVID-19 pandemic, women have shown better survival rates than men but may be more prone to long-term health problems.[4] Pulling apart the biology of the disease, and remembering these sex-related differences, we can see why this might be. COVID has a dramatic effect on the heart, with greater clotting in the blood, as well as damage to the lining of the blood vessels that encourages clots to attach. These clots can block the vessels and trigger a heart attack. We know that existing heart disease or conditions that underlie heart disease—hypertension, atherosclerosis, obesity—predict who is most likely to die in intensive care. Men are more prone to have heart disease and many of these underlying conditions. Women are less likely to become infected, because of their more active immune systems, and less likely to have the existing cardiac problems. But there can be a second wave of damage from COVID-19 that comes later, when the immune system flares up and overreacts. Now the body is attacking itself and wreaking havoc on the already weakened heart and lungs. (Drugs like dexamethasone, found to improve survival in intensive care, work to damp down this extreme reaction.) "Long COVID" is the result of this catastrophic damage, characterized by months of fatigue, breathlessness, and heart problems. Early results showed that women have been most affected by long COVID.[5] We can see that their more vigorous immune response might be the reason for this self-damage, just as in autoimmune diseases.[6] Interestingly, while COVID-19 is most dangerous for the elderly, long COVID seems to happen more in middle-aged women. We might hypothesize that the effects of age that damp down the immune system reduce the risk as women get older.

So, biological sex is crucial to the heart. Some of the differences in heart disease between men and women are baked into our genes. But it is the lifelong ebb and flow of hormones that carries out the genetic instructions and—in the process of scripting our reproductive roles—widens the gap between the sexes.

STEWING IN THE HORMONE SOUP

Scientists often get laughed at, and not just for their taste in clothes. We do experiments that seem pointless because "everyone knows" a particular fact and so it's "just common sense." This phenomenon is rife in the world of sex, gender, and hormones. It's clear that up to about the age of 60 or so, men are more susceptible to heart disease than women (figure 8.1). So it must be the hormones, right? Viewed simplistically, men have testosterone (bad for the heart) and women have estrogen (good for the heart), and when the estrogen wears off at menopause women become equally at risk. There is some truth here: both hormones have strong effects on blood vessels and blood clotting, so there are many reasons to expect them to affect heart disease. Risk goes up at puberty for sudden death from hypertrophic cardiomyopathy, and Takotsubo cardiomyopathy is much more likely after menopause. It all makes sense. Except when you dig down, there are a lot of unanswered questions.

First, testosterone and estrogen are not uniquely linked to the sexes. Testosterone is continually converted in the body to estrogen (or to be precise, estradiol, since estrogen is a group of compounds). Adult men will have about 20 percent of the estradiol levels of an adult, premenopausal woman. Ovaries produce both estrogen and testosterone, and the testosterone level in a young woman can also be as much as 20 percent of the level of the average man. Both hormones drop with age in both sexes. Testosterone can be a very useful addition to hormonal therapy for women at the menopause, while increased estrogen-to-testosterone ratios in aging men can contribute to low libido and middle-aged spread.

Given this blurring of the biological divide, we should not be surprised that there can be fluidity of perceived gender in both men and women. More people are now recognizing that they are not in tune with their birth sex and are seeking to undergo gender reassignment. Often, the first stage is to suppress their existing hormones and treat them with those of the desired sex. Because heart disease is so clearly linked to gender it has been important to track the effects of this hormone switch. The data coming out are at an early stage but throwing up some very interesting findings.[7]

The initial surprise is that testosterone is not the bad guy we thought. Females who were transitioning to males (F-to-M) and having testosterone supplements showed very few adverse side effects on the heart. Of course, they must be compared with both untreated women of the same age as well as untreated men, which makes it more complicated. Reassuringly, they didn't adopt the male pattern of coronary heart disease or myocardial infarction. They had a slightly increased risk of high blood pressure but the cholesterol markers in the blood were variable and some even improved. The second surprise is that the males who were transitioning to females (M-to-F) did not benefit from the female pattern of heart disease protection. In some (but not all) studies they had an increased risk of myocardial infarction and hypertension. The risk was even greater for thrombosis and stroke, linked to increased blood clotting activity. It is as if the known effects of estrogen to increase blood clotting have been added to the lifetime male risk of cardiovascular disease in the transitioning M-to-F subjects.

But maybe this wasn't so surprising after all. Doctors have been trying to supplement hormones for many years, in men or women with hormone deficiencies, and continually have been frustrated when their simplistic assumptions turn out to be wrong. Testosterone therapy has always been controversial because of the link between male gender and heart disease and so there has been a great deal of scrutiny of the treatment. It seems that there is no problem when restoring normal levels of testosterone in patients when we are trying to treat specific diseases that reduce testosterone production.

For men suffering symptoms because of age-related testosterone loss, it's not quite so simple.[8] Cardiovascular disease is more common as testosterone drops, but is this a causal effect or just a general marker of ill health? Reducing testosterone to treat prostate cancer hasn't shown any increase in heart disease, so could the age-related drop be unimportant? Giving testosterone to older men, or those with heart disease, seems to have effects that might be beneficial—reduced obesity and diabetes plus greater exercise capacity. Yet some studies have shown that elderly patients may have an increased heart disease risk after treatment. Unhelpfully, two large analyses of many studies have come to opposite conclusions. It's all very confusing.

MIXING THE HRT COCKTAIL

Our efforts to preserve the cardiac health of women by treating
the loss of estrogens at menopause have been equally bewil-
dering and frustrating.[9] Women can suffer many debilitating
symptoms as hormone levels start to drop—hot flashes, tiredness, and insomnia—and it is these that drive them to look for
help, rather than a concern for heart fitness. At first it seemed
that the heart did benefit as well from hormone replacement
therapy (HRT), just comparing the treated women with the
general female population around menopause. But part of
this effect was simply that women who sought out HRT also
looked after their health in other ways: they were generally
better off and better informed. When the first large, controlled study was done, the results were a shock—the risk of
heart disease seemed to increase in the treated women! Not
only that, but breast cancer risk was increased. True, there is
also a beneficial effect on bone density (and so reduced frac-
tures) and a reduction of colon cancer, but the overall effect is
that the popularity of HRT has plunged.

This legacy of fear about HRT has been hard to shake, but
scientists are arguing for a more balanced view.[10] More trials
have been done, with different age ranges of women and different mixtures of hormones. In healthy younger women in
early menopause (below age 60, or within 10 years of menopause) mortality was reduced by 40 percent during treatment.
Importantly, this benefit persisted after the treatment and for
18 years of follow-up (the effect on bone density was also
maintained for this time). Older women, further from menopause, and with other health conditions were the ones who
did badly. Women with the worst symptoms of menopause
are more likely to have a risk of developing cardiovascular

disease later and these women got the most benefit from HRT on heart health. Tailoring the dose and mixture of estrogens and progestogens to the individual woman can also reduce side effects and improve the benefit for the heart.

The risk of breast cancer does remain, but it should be put into context. In 1,000 women between ages 45 and 79, there will be 23 diagnosed with breast cancer. Treatment for breast cancer is excellent, with over 90 percent of patients now surviving for over five years after diagnosis. Five years of HRT is linked with six extra diagnoses (not deaths) and this compares to three extra for smoking, six extra for moderate alcohol intake, and four to seventeen for overweight/obesity. Regular physical activity can produce a *reduction* of 10 diagnoses. Doctors and scientists are now arguing that the health benefits of a short course of HRT soon after menopause outweigh the potential risks of a breast cancer diagnosis, and that many women are being unfairly deprived of this important treatment.

THE INVISIBLE WOMAN

Women do worse after a heart attack; this is now clear from many studies. They are less likely to be diagnosed correctly, less likely to get the best surgical treatment, and less likely to be discharged with the optimal set of drugs. None of this is excusable, but is it understandable? Let's look at the justifications given by doctors one by one and see how they hold up. First excuse commonly offered: women don't get as much heart disease and so seeing a woman with a heart attack is "unexpected." It's true that the heart disease rate is lower in women, as we have said, but it is far from a rare event. For every 10 men with heart disease a doctor can expect to see three to five young women but seven to eight older ones. Compare this with meningitis, where a general practitioner may see only one or two cases in her entire career. Considering all heart disease types over an individual's complete lifespan, around 21 percent of women die from heart disease compared to 24 percent of men. For a clinician, seeing a woman with heart disease in the emergency room cannot be called "unexpected" by any stretch of the imagination.

A second popular excuse: women's symptoms are strange and unpredictable. The reality is that heart attack symptoms can be variable and there is a great deal of overlap between the sexes in what they experience. The classic symptom is crushing chest pain, often radiating up the arms and to the jaw. This is the most common symptom in both men and women, although women are more likely than men to experience it in their back. Both sexes may feel sick, sweaty, or light-headed, and this can be perceived as an anxiety attack in women especially. Shortness of breath and tiredness are again common, and it is a bit more likely that women will have the symptoms of breathlessness, fatigue, or nausea when they come to the emergency room. As we have established, there will be a significant number of women coming to the hospital with a heart attack, so there should be no excuse for remaining ignorant of this range of symptoms.

Once a heart attack is suspected in a patient, the standards and guidelines for treatment are very well defined. Doctors should be recognizing heart disease in women and they should be giving them the optimal standard of care. However, this is not happening. Clinicians are less likely to stick to the guidelines when treating women, sending them home with painkillers rather than the armory of therapeutics we have now.[11] Women are less likely to receive the gold-standard treatment

where the blood vessels are opened to restore the blood flow
using catheters. One study of more than 100,000 hospital
patients saw men had 20 percent more of these reperfusion
treatments than women and were nearly twice as likely to survive while in the hospital.[12] Even when women do get treated,
there is not as much haste in doing so. Time from first contact
with a doctor to reaching the catheter laboratory for reperfusion is vital: for every five-minute delay there is a 5 percent
increase in risk of death. The study found that women were
moved to the catheter laboratory significantly less promptly
than men, and this contributed to the higher death rate. However, the most shocking statistic was that this only happened if
the doctor was male. Why should this be?

THE HYSTERICAL WOMAN

Clinical cardiology traditionally has been a predominantly
male occupation: sometimes called "boys and toys" because
of the many devices that can be implanted in the heart, pur-
portedly attracting male clinicians to the discipline. The UK
Athena gender equality scheme was operating for about 10
years in my own university and made many adjustments to
reduce bias in hiring and promotion practices. By 2020 we had
raised the number of female *science* professors in cardiology to
be about equal with the number of males. Female *clinical* professor numbers, however, remained stubbornly at 10 percent
of the number of males in our associated hospital, where the
Athena scheme was largely ignored. In the United States more
than 50 percent of medical school students are women but this
figure drops to only 4.5 percent for the practicing interventional cardiologists (who are the ones using catheters to treat

heart attacks).[13] This difference seems to be crucial in the poor treatment of women who come in with cardiac symptoms.

The biggest study on physician gender and treatment came from the experience of 1.3 million Florida residents who had been admitted to the hospital for a heart attack.[14] Survival rates were two to three times higher for female patients treated by female physicians compared with female patients treated by male physicians. Male physicians who had a good prior experience of treating women did improve their success rate— there was a measurable increase in survival with every new female patient they saw. Even more interesting, the number of women clinicians in the team made a big difference to the men they worked with. A higher proportion of female doctors improved both the success of the team in general and the competence of men in the team for treating women. The study concluded that the best way to help female patients was to have a gender-balanced team, rather than waiting for individual male doctors to gain experience at the expense of their early failures.

On a side note, one clue to understanding whether an effect is a result of bias for women is to observe whether the same is seen for other disadvantaged groups. The same phenomenon for doctor/patient matching occurs for race, with patients in minority groups doing better with a physician of the same race, or in a team with a good proportion of minority doctors. Use of health care resources and satisfaction with the outcome both rise when there is good matching.[15] This is part of a much wider appreciation that there are numerous inequalities in medical care by race in both the United States and UK.[16] It is not hard to predict the pattern for minority women, who are doubly disadvantaged in terms of health care.[17]

What is it about female patients that makes the male doctors treat them differently? What behaviors or characteristics
trigger this response in the clinician? This is where the difference between sex and gender really plays a part. Each of
us, independent of our biological sex, has a range of gendered
attributes that are traditionally thought of as male or female
and, importantly, that might be valued differently if displayed
by a man or a woman. Are you shy, gentle, and compassionate
or assertive, risk taking, and individualistic? There is a test you
might like to try called the BEM Inventory score that assesses
how "male" or "female" your behavior is—almost all of us
will fall somewhere between the two extremes.[18] Our home
circumstances also affect how we are seen: factors like being
the primary wage earner; having a high income; or doing
most of the housework. All of these add up to how "male"
or "female" we appear. When gender and biological sex were
compared for how they influenced treatment, it was the perceived gender—the strength of the "female" score compared
to the "male"—that made the difference in treatment and
outcome.[19] For example, "female" patients (men or women)
were more than four times as likely to return to the hospital
with recurrent symptoms after being discharged. Essentially,
behaving in a manner perceived as traditionally female down-
grades you in the eyes of a male physician—there is a higher
chance that your distress will be seen as overblown, inaccurate,
or "hysterical."

Uncontrollable emotional excess has long been associated
with women and has frequently been classified as a disease
of either the body or mind. The Greeks termed it "hysteria"
(wandering womb; *hystera* is the word for "womb") and only at
the time of Freud was the same behavior pattern recognized in
men. In her book *Sex Matters* the physician Alyson McGregor

describes how women who are in pain often have trouble convincing the doctor treating them of how serious that pain is.[20] The more they protest and try to convince the physician, the more their behavior is perceived as "hysterical." Women from more demonstrative cultures have a particularly hard time. If they have grown up always encouraged to be very vocal about their emotions, then this can work against them in the emergency room. As McGregor says, the best thing you can do as a woman is to bring a man with you to explain.

Finally, going back to our Takotsubo syndrome sufferers, we can see that they are at the center of the storm of sex and gender differences. Their symptoms are a direct result of the biological adrenaline surge but are often precipitated by severe emotional distress. The root is emotion but the effect on the heart is physical. These women above all have a big problem with being believed—they are lucky in a sense if they get their heart imaged, because at least there will be an observable physical sign. How many more go home with no diagnosis? At the hospital where my colleague specializes in Takotsubo syndrome, we have a loyal and talented group of women who have recovered from the disease, but who continue to help us with our studies. These women are simply grateful to be believed. They come from all walks of life, many are highly accomplished in their work or have a wide range of responsibilities. In all other respects, they are used to having their voice heard and respected. They have often arrived with my clinical colleague after being passed around from one physician to the next, or even different hospitals, without ever getting a proper diagnosis or a clear explanation for what they experienced. Now their friends and family can understand that this is a real disease and not a figment of their imagination. At last, it is clear they are not just "hysterical."

In this chapter and those preceding, I have tried to show how the new science is unlocking the secrets of the heart. We have discovered many threats, old and new, which makes us appreciate the incredible robustness of our one-and-only hearts. We have seen how the present therapies are working, as well as when and why they fail. Now we will journey to the outer reaches of science, where the emergence of bionics, cyborgs, edited genomes, and artificial cardiac muscle bring us to the edge of science fiction. I will help you understand how scientists must keep upping their game to match what evolution has created in the exquisite machine of the heart.

THE MECHANICAL HEART

In the film *Up in the Air* George Clooney's character, advising on airport time-saving hacks, counsels that one should never get behind old people in the security line as "their bodies are littered with hidden metal." It is true that implantation of metal or plastic devices has stealthily become commonplace in modern medicine. Right at the top of the list for this metal load, along with joints, is the heart. Stents, pacemakers, artificial valves, and defibrillators are common; then the new generation of wireless recording devices; and finally, the big guns of partial and full artificial hearts. We are used to heart function being taken over in the operating room with bypass machines and other devices. Rapid advances in technology have either made all these possible or dramatically improved their function. The same innovations that make your smartphone better, your car lighter, and your TV clearer have been harnessed for the benefit of your health. I will start with showing you the relatively straightforward (but still highly complex) technological control of rate and rhythm then work up to the huge challenge of replacement of the whole heart with a mechanical one, an Everest we have still to conquer. To put this in context, the project to develop the first total artificial heart was started at the same time as the one to reach the moon, and yet today we still do not have a replacement organ to offer the patient in end-stage heart failure.

KEEPING UP THE RHYTHM

Pacemakers are one of the most frequently used devices, with 200,000 pacemakers implanted annually in the United States.[1] They give back a stable rhythm to the heart when its own internal pacemaker starts to malfunction or when the cardiac muscle fails to respond to its signals. The aging of the population and increased realization of how many patients might benefit from pacemakers means their use is growing. Pacemakers are surgically implanted just below the collarbone and electrical leads are fed into the heart through a vein, then anchored in place. The pulses from the pacemaker then trigger the beats at the right interval—especially useful for people who have an abnormally low or irregular heartbeat. The first pacemakers could only pulse at a fixed rate, but now accelerometers like those in smartphones or wrist devices enable pacemakers to speed up when your walking pace increases.[2]

One of the biggest problems with any artificial device implanted in the body is infection. Pacemakers are contained completely within the body, so at least there is no infection from outside to contend with. However, the long leads from the main body of the device into the heart are a potential breeding ground for bacteria, and infection through the blood vessels is a major source of complications. A new generation of leadless pacemakers is looking to solve this.[3] The miniaturization of components allows these tiny devices, around the length of a jellybean, to be directly implanted in the heart (figure 9.1). There is a natural response of the body to grow tissue around any nonbiological material, called the "foreign body reaction." This encases the tiny pacemaker and shields it from the blood supply, reducing the chances of infection and of blood clotting, a further danger from implanted materials.

Figure 9.1
The changing size of pacemakers

The power requirement, another continual challenge for devices, is reduced with miniaturization and a single leadless pacemaker can last for 10 to 15 years without changing.[4]

Control of rate has been beneficial in other ways for the failing heart. It is important that the left side of the heart, which propels the blood around the body at high pressure, synchronizes precisely with the right side, which sends blood through the lungs for oxygenation. Scientists observed that the two sides of the heart could start to become slightly out of time with each other when the heart is severely enlarged in heart failure, and that this was often due to slowing of the electrical impulse movement across the left ventricle. Cardiac resynchronization therapy (CRT) uses a pacemaker to bring the two sides back in harmony and has been successful in improving the output of the failing heart, increasing survival rates, and relieving some of the symptoms of breathlessness and fatigue.

Making CRT leadless is much more of a challenge since the wires must span both the right and left ventricles and communicate with each other. New pacemakers are being developed with communication modules, so that they can coordinate the pulses. Two or three leadless pacemakers can then be positioned around the heart and work together to time the signals optimally. Their signals must not interfere with the natural electrical activity of the heart, which means that the more usual radio frequency cannot be used. Instead, very high frequency signals are emitted, which do not interfere with the heart and are also resistant to interference themselves. These signals travel at almost the speed of light and each pulse contains multiple pieces of information.[5] The first small-scale studies started in 2019, with one study of 100 patients showing successful control of heart rhythm.[6]

The most devastating threat to the rhythm of the heart is ventricular fibrillation, the wriggling "bag of worms" heart that precipitates sudden cardiac death. Defibrillators are now common in workplaces, railway stations, and large stores, but the four-minute window before the brain is irreversibly starved of oxygen does not give much time to run and get the kit. Defibrillators are now being implanted in the hearts of patients at risk, to sense abnormal rhythms and deliver an internal shock directly to the heart muscle. Conventional implantable cardioverter defibrillators (ICDs) have a high-energy battery and coiled leads (capacitors) that are charged to high voltage, enabling the delivery of high-energy shocks. Unfortunately, the miniaturization of batteries is not yet good enough for these very high-power requirements, but the leads themselves have been reduced in size with new developments in technology.

Like the CRT, the idea is being tried for ICDs with a leadless battery in the ventricle, and a separate device to detect the abnormal rhythms and trigger the shock.[7] This time researchers are investigating radio frequency near-field effects, the kind of technology that enables your contactless bank card to work, to make the connection. It is vitally important that the detection and communications systems are accurate and efficient, although in fact only a relatively small number of the 75,000 ICDs implanted in the United States annually will finally deliver a lifesaving shock.[8] This accuracy is important to catch the rare but dangerous arrhythmia, but that is not the only consideration. Each shock feels like the kick in the chest, as you might get from an external defibrillator. Extra unnecessary shocks can ruin the life of a patient if they become too frequent and so the device must be "intelligent" enough to ignore the more minor events or general electrical noise. Algorithms that can learn from experience and innovations in communications technology are driving forward huge advances in cardiac rhythm devices.

MESSAGES FROM THE HEART

If we can communicate between devices within the heart, can we communicate from the heart to outside the body? The game-changer for pacemakers, CRT, and ICDs has been leadless devices that separate the sensor from the stimulator and send signals between them. It would be enormously useful to analyze the information from the sensor to understand the development of disease and its response to therapies. Remote monitoring sensors (like Fitbits, discussed earlier) that can send back signals about heart rate or blood pressure are very valuable, but they don't give direct measurements from the heart itself. To harvest information from internal sensors would be a big step forward.

Loop recorders are the first fruits of this insight, miniaturized sensors that are implanted for a fixed amount of time to transmit and store electrical data and then taken out again.

These have been a godsend for patients with undiagnosed arrhythmia, often identified after episodes of fainting, who would usually have had to wear a Holter monitor. A Holter is a portable version of the static ECG/EKG in the hospital, with leads attached to your chest for continuous recording of heart rhythm. Because it can record for longer than the few minutes of the static one, it is more likely to catch an arrhythmia. Even though the design of the Holter monitor has been improved by miniaturization, so that it is about the same size as a deck of cards, it can still be limiting to wear around your neck for longer than a few days. And even three or four days can be too short for patients who have more random or intermittent arrhythmia.

Miniaturized loop recorders are around an inch long and can be implanted under the skin above your heart in only a few minutes.[9] The sensor monitors your heart rhythm continuously, storing a few minutes of data at a time. If there is no arrhythmia in this period then the recording is erased, and new data is stored. If an arrhythmia is detected then an alarm is triggered, and the information sent via an internet connection to the clinician for diagnosis and treatment. If you feel unwell or sense something wrong, you can press a button to record the time, linking disturbances on the loop recorder to real-time events. Loop recorders can stay in place for several years or be taken out (leaving only a tiny scar) when not needed or the battery is dead.

More ambitious still are the very new devices that aim to send out complex information about the contractile function of the heart. Heart muscle contraction's chief function is to generate increased pressure within the ventricles to force out blood with each beat. Failing hearts produce only slow and weak changes of pressure when contracting and, during the relaxation part of the beat, the pressure remains higher than normal because there is excess blood that has not been ejected. These faults translate into the symptoms of fatigue, as well as water retention around the lungs and in the limbs. Heart failure patients can stay stable for years but then have acute periods where these symptoms get worse and they must come into hospital for adjustment of their medication. This is a huge burden both for the patient and for the health care system.

Of note, there is often a series of warning signs before symptoms develop, when the pressure changes in the heart begin to deteriorate. The latest internal monitor has a pressure sensor that is implanted via a catheter, to avoid open-heart surgery, and is attached to the wall between the two atria.[10] Power is supplied and signals relayed by an external pack, worn on an over-the-shoulder strap. All the data is stored on a cloud server and can be studied by the clinician, who will then contact the patient or their caregiver to change the drug regimen. Before the patient even knows they are worsening, the system will intervene to prevent the oncoming episode and save a hospital stay.

So far, all the devices I've talked about are aimed at sensing or correcting specific aspects of heart failure or rhythm abnormalities. However, the Holy Grail is a complete replacement for the failing heart. With all the advances in technology and materials we have developed in more than 50 years, have we moved closer to this goal?

THE IMPOSSIBLE MACHINE

Nothing shows more clearly the perfect engineering of the heart than our own failed attempts to imitate it. This history of the total artificial heart is punctuated with both brilliant innovation and continual clinical failure. In 1962, John F. Kennedy challenged the scientific community to land a man on the moon and return him safely to Earth by the end of the decade. In 1964, cardiovascular surgeon Michael DeBakey persuaded President Lyndon B. Johnson to fund a program to develop the first functional self-contained artificial heart, launching a race to successfully make one before the moon landing. In 1969 both aims were apparently achieved, with the Texas Heart Institute implanting the first total artificial heart just three months before the launch of Apollo 11. However, while the moon landings have led to the Space Shuttle, Mars Rover, and International Space Station, and (despite a long lull) the newest aims to develop a moonbase to bring us to Mars, a reliable off-the-shelf total artificial heart is still just out of reach.

At the outset, the artificial heart was aimed to be a lifetime replacement for the failing organ. This was a high bar to reach, since the first design had an external compressor with an air line through the skin into the patient's body. Compressed air inflated and deflated Dacron pouches or sacs, which collapsed and expanded to displace blood from a surrounding sac. While having the compressor outside the body was useful, since the mechanical parts (which were most susceptible to wear) could be easily replaced, it would make for a bulky piece of equipment to be wheeled about with the patient. It was difficult to see how this could be given to a patient and expect them to live an even partly normal life for many years.

However, the history of the artificial heart is also intertwined with that of the heart transplant. This was again only a hopeful dream in the early 1960s, but by 1967 cardiac surgeon Christian Baarnard in Capetown performed the first successful transplant. Now, the purpose of these first artificial hearts was changed. They did not need to be suitable for a lifetime; their purpose was to keep the patient alive until a transplant donor could be found. As with many highly experimental therapies, the first case was done on a patient who had run out of options. A 47-year-old man was being operated on to repair a huge aneurysm of the left ventricle that had thinned and swollen the heart wall. He was being supported by a heart-lung machine, which bypassed the heart and kept blood flowing through the body. However, he could not be weaned from the machine at the end of the operation as his heart was too weak. He desperately needed a transplant. Denton Cooley, DeBakey's associate, offered him the new experimental total artificial heart and he accepted. The patient was kept stable with the new device for 64 hours until a matching donor heart was found and then transplanted.[11]

This seemed at first a triumph for the total artificial heart, but tragically the patient died 32 hours later from sepsis. Not only that but the device had damaged both the blood and the kidneys, and the walls of the expandable sacs were coated with blood clots. This heralded a series of problems that would continue to thwart the scientists and engineers wrestling with this procedure. Infections and sepsis are a continual challenge to any device where there is a wire that must permanently cross the skin. Devices that move the blood will alter its composition and the foreign surfaces will cause the blood to clot, resulting in strokes and blood breakdown. The first Jarvik heart, one of the next iterations, was implanted in five patients and one lived for 620 days. But two of the patients had severe strokes, and eventually all died of either sepsis or blood problems.

Heart transplantation also had a shaky start, with Baarnard's first patient dying after only 18 days. The first patient in the UK, whose transplantation was performed by cardio-thoracic surgeon Donald Ross at London's National Heart Hospital, survived for only 45 days, and the general success rate remained disappointing. The problem here was not the mechanics of the operation or the initial performance of the new heart. It was the mismatch of the immune system of the recipient to that of the donor heart. Even though the donor heart is matched as closely as possible to the patient with the major tissue types, the immune system must be suppressed to stop the heart being rejected. Drugs to suppress the immune system were not very sophisticated in the early days, but the development of ciclosporin in the early 1980s produced a revolution in immunosuppression that dramatically improved the success of heart transplantation. Now, it is a victim of its own success, with many more people in need of a transplant than there are donors. Only about 200 transplants are carried out in the UK each year despite more than 750,000 living with heart failure, and similar figures are seen worldwide. To fill this gap, scientists have been genetically modifying pigs to make their hearts compatible with the human immune system so that they can be transplanted to patients without being rejected. This has proved very complex and challenging, but first clinical transplants have started in 2022.

The success of heart transplantation, however, had reinvigorated the search for the total artificial heart, with the more achievable goal of keeping the patient alive until a donor is found, or "bridge to transplant" as it is called. For decades, the artificial heart technologies have improved through changes to more biocompatible materials, better valve design, and more efficient handling of blood flow. Successes have been achieved: one study saw 80 percent of patients on the artificial hearts surviving for over a year, and some for 6 years.[12] The longest time a patient was supported to transplant was 1,373 days.[13] But severe infectious complications were still common, and the goal of a complete "destination" therapy for artificial hearts was still a distant dream.

Meanwhile, the urgent need to bridge to transplant had taken the technology in another direction. Rather than replacing the failing heart completely, the idea was to support it by assisting the blood flow. The ventricular assist device, or VAD, took blood out of the ventricle of the heart by a completely different route and pushed it into the aorta at high pressure (figure 9.2). This added to the blood being ejected

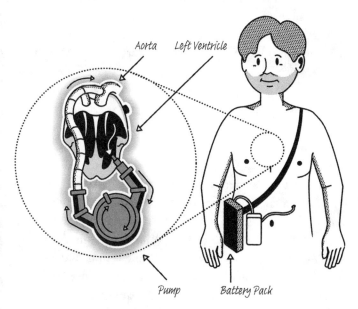

Figure 9.2
Left ventricular assist device: blood is pumped out from the apex of the left ventricle and expelled into the aorta to supplement blood ejected from the heart

from the heart and thereby magnified the effective cardiac output. It also solved another problem encountered by the engineers of total artificial hearts—how to balance the right and left heart-blood flow. The amount of blood circulating in the left ventricle/body loop must be very close to that in the right ventricle/lung loop. With 100,000 beats a day, even a teaspoon of difference at each beat would add up to 500 liters of blood in the wrong place. The heart has evolved complex biological mechanisms to make sure this does not happen, but the engineers were having huge battles to try to do the same with feedback systems. For VADs, either the right (or more usually) the left ventricle can be independently supported, taking this problem away.

Left ventricular assist devices, or LVADs, have produced a revolution in care for end-stage heart failure. More than 15,000 LVADs have now been implanted worldwide, and around a third of patients with end-stage heart failure are now supported on LVADs. The intention is usually to bridge the patients to transplant, but in fact the shortage of donor hearts means that patients can often stay on LVAD support for years. Survival rates of over 50 percent are seen at seven years, and there are reports of patients living up to 13 years on these devices. LVADs have therefore become by default a ther-apy in themselves.[14] Again, technology has progressed, with newer LVADs performing better. A breakthrough idea was to stop imitating the heart, with its pulsing action, and move to constant flow of blood. Rotating paddles (impellers) push the blood along in a continuous motion, creating a smooth unbro-ken stream. This has the curious side effect of creating a patient without a pulse, which can be disconcerting for the unsuspect-ing physician as well as producing some unwanted side effects as the body adapts to the new flow. External battery packs are still an inconvenience and a source of infection, but systems are being developed that transfer energy transcutaneously (across the skin) based on induction (like domestic induction stoves). The LVAD units would still need a small, implanted battery in case of a temporary device failure—and it has been known for external battery packs to be snatched from patients by handbag thieves!

The search for a completely implantable total artificial heart continues. Trying to develop external transcutaneous units to fully power the demands of the heart is the biggest barrier. Specifications for a total artificial heart require it to pump eight liters per minute of blood against a blood pressure of 110 mmHg. (The biological power storage molecule adenosine triphosphate [ATP] would be needed in quantities greater than half your body weight per day to power your own heart to do that, if ATP were not continually renewed in cells.) Compressors have been miniaturized to be more portable, but it has been a struggle to make them completely implantable. Here it seems that the VAD technology may hold a solution, dispensing with compressors altogether and using instead the impeller devices, with dual right and left VAD working together.[15] Solutions seem tantalizingly close, but no one is anticipating an easy ride. The many failures over the years have certainly produced in scientists a humility and awe for the natural engineering of the heart.

THE GIFT OF LIFE

In a final twist to the story, the endeavors of the engineers and those of the transplant surgeons have come together. Of course, transplants would not even be possible without the sophisticated heart bypass machines used to support the body while swapping the failing heart for the donated organ. During a transplant operation it is unnerving to see the patient on bypass lying with an empty chest after the old heart has been removed and before the donor heart has been implanted. Even more complex, bypass systems such as ECMO (extra-corporeal membrane oxygenation) give extended replacement for both heart and lung function in extreme situations and can be maintained for weeks. People who need ECMO have a severe and life-threatening illness that stops their heart or lungs from working properly. For example, ECMO is used during life-threatening episodes such as severe lung damage from infection, or shock after a massive heart attack.

Ultimately, the bottleneck is the number of hearts available for donation, which is far below that needed for all the waiting patients. To compare the number of donor hearts with the patients on waiting lists is misleading, because only the most urgent cases even make it to the waiting list. Typically, a donor might be a young person who has been the victim of a trau-matic accident, such as a motorcycle crash, and suffered a head injury. Brain stem death has occurred, but the heart is still beating. The clinician who must ask the family for an organ donation at the time of their worst imaginable pain has a ter-rible conflict, for they understand this is the end of hope. Campaigns to encourage people to carry donor cards, or the newer legislation that assumes donation unless the family opts out, can lift to some extent the burden of choice. Highly orga-nized networks, where the hearts can be airlifted to distant destinations within the six-hour window of usability, have meant that few donor organs are wasted.

But even with these measures the number of donors is not increasing fast enough, in part because increasing motor vehi-cle safety has reduced traumatic injury and death. In addition, fewer donor organs are in peak condition because some populations' poorer fitness has potentially reduced the quality of the available hearts. Donors are now slightly older on average and, in the United States, slightly heavier. In Europe, donations increasingly are coming from people dying with stroke or brain hemorrhage and less from those suffering head trauma deaths. In the United States, head trauma is still the most common cause of death in donors, possibly because of continuing gun incidents, and deaths from drug overdose are also a large contributor.[16] Fewer donors are smoking but there are more with diabetes or a history of hypertension. However, while donor health is decreasing, the hearts that are being accepted as suitable are in better condition. Part of this may be due to some very recent technology—the Organ Care System.

In my laboratory, we have used a clever trick to keep hearts functioning outside the body, first invented by Oscar Langendorff in 1895. Normally, blood would be ejected from the left ventricle into the body through the aorta, via the aortic valve. Just outside the aortic valve is the entrance to the coronary blood vessels, which supply the heart muscle itself. Passing fluid backward through the aorta closes the aortic valve and forces the fluid to go into the coronary vessels. So, if we connect a tube to the aorta of the dissected heart, we can pass an oxygenated salt and glucose solution through the coronary vessels this way. This keeps oxygen and nutrients flowing through the heart muscle and preserves its function.

The Organ Care System uses this trick to perfuse the donor hearts while they are in transit to the patient waiting for transplant.[17] Continuous warm perfusion with oxygen and nutrients increases the time for the heart to remain in good enough condition for transplant, which in turn increases the range it can travel and the number of eligible people who might benefit.

We can also see the condition of the heart better and reject hearts that are substandard. This increases the chances that the transplant will be a long-lasting success. Sometimes, the heart even improves on the Organ Care System. Adrenaline is often released in huge quantities by head injury and traumatic death, and this can give a kind of Takotsubo-like syndrome, depressing the function of the heart muscle. As with Takotsubo, this is reversible: over time on the Organ Care System the function starts to recover. Rescue and preservation of pediatric donor hearts is expanding the pool of available hearts for young children.[18] Even hearts that have stopped before the surgeons were able to retrieve them can be connected to the system and assessed, and sometimes rescued.[19] Every new heart narrows the gap between the desperate patient in the last stages of heart failure and a life-saving transplant.

So, engineering alone cannot yet replace the heart, but it can be some help to the patient. Where can science take us now for the next advance, to finally give us back the functioning hearts we need?

CAN STEM CELLS HELP US GROW A NEW HEART?

Stem cells are always in the news, with excited scientists hailing yet another breakthrough. So why don't we all have a spare heart in the fridge for when ours starts to wear out? In this chapter I'm going to separate out the hype from the reality and show you where we are on this journey to replaceable organs.

First, it is hard to overstate just how astonishing the science of stem cells really is! One remarkable but little-known fact is that stem cells that cross into a mother's body from her developing baby will stay with her for life. After pregnancy with a male fetus, we can pick these cells out in the mother because half of their DNA comes from the father of the baby. These fetal stem cells have been detected in women in their 70s and 80s, decades after their pregnancies.[1] I find it amazing and touching that my daughter's cells will be with me for life. We have watched, fascinated, as these stem cells rush to the area of a wound or surgery. In a mouse who is having babies that carry a harmless bioluminescent protein, we can use a special camera to view the fetus from outside the body. Cells detach from the baby mice and move into the mother's body. If a tiny scratch is made on the mother's ear, then the stem cells rush to the site of injury.[2] In the same way after breast cancer surgery, women have been found to have their children's stem cells clustered around the wound.[3] We strongly suspect that they are involved in some repair function in people, and experiments in the mice back this up. Cells from our children may be helping to keep us healthy—repayment for all those sleepless nights! Perhaps this is evolution's way of getting your mother to keep doing your laundry.

As we studied the amazing biology of the stem cells, we have unpacked the secrets of their rejuvenating power. We have worked out how the cells can grow and divide vigorously, and how they then choose what cell type to finally become. Armed with this information we can now create our own. From only a scraping of your skin or a few blood cells we can now make a dish of stem cells, by turning on the genes that keep them in the pluripotent stem-like state. By changing the cell culture conditions again, we can remind these stem cells how to become like those in the heart, and they can then go on to produce cardiomyocytes, blood vessel cells, and the structural fibroblasts.

What we really want is to use stem cells to combat disease. Bone marrow transplants give us hope for stem cell therapy, because we have been successfully performing these operations for more than 50 years. Before the transplant, the whole existing bone marrow of the patient, with its defective or cancerous cells, is destroyed by radiation or chemical treatment. A volunteer donor has a sample of their healthy bone marrow taken from a hip bone, and this is injected into the patient. The new bone marrow from the donor contains a tiny group of stem cells. These take up residence in their new home, expand in number, and regrow the whole of the patient's blood system with healthy functioning cells. It is fascinating to think that this patient now has two sets of DNA—their own and the donor's. Like the mothers with the fetal stem cells, the patient is a chimera: that is, they are a mixture of cells from different individuals.

The portents for stem cell therapy are good and we have made incredible progress even in the last five years. But stem cell biology is complex and is unfolding day by day. Some of the persisting problems involve understanding the stem cells, but some now are frankly a matter of logistics. How can we get these cells past the hypervigilant immune system? How can we engineer a construct that the heart will accept? Who owns stem cells that have been made from your skin? What is the business model for companies to sell your own cells back to you? The questions start with the biology, but they don't end there. The road has been long, exciting, and sometimes frustrating.

THE HEALING BODY—STEM CELLS AND YOU

What is a stem cell? It is a cell that can grow and expand continuously but has not yet committed to a being a specialized organ type—this is called the "undifferentiated" state. A stem cell is also capable of differentiating, or switching into a specific type (for example, skin, heart, kidney) when it gets a certain signal from the body. In the very early embryo, just 10 days after fertilization, there is a tiny bundle of cells from which every organ in the adult body will grow. These are the controversial embryonic stem cells and it is from them that we have gained much of our knowledge about stem cell biology (figure 10.1). They are pluripotent, which means that any single cell can become any one of the cell types in the body. There is only one cell type more powerful (or totipotent) and that is the fertilized egg: this will produce the complete and organized body of the baby plus the placenta.

In your body now there are many stem cells that have a more limited range: these are the multipotent stem cells. They

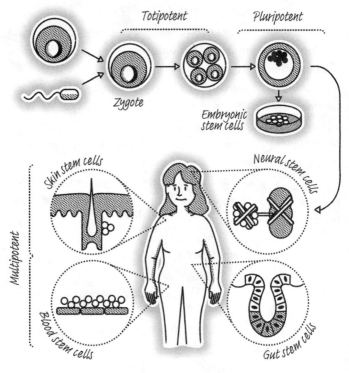

Figure 10.1
Stem cells within the body. The fertilized egg (or zygote) is totipotent and can produce a complete baby. Embryonic stem cells can produce all cell types of the adult body except placenta. Multipotent stem cells can produce either one type or a small range of cells for organ repair.

usually live in the tissues and their job is the day-to-day repair of our organs as they wear out or are damaged (figure 10.1).[4] These stem cells give the body its amazing power to heal and adapt. Skin is the exemplar for repair; we seldom consider how remarkable it is that the gash on our hand first develops a scab and then closes between one day and the next. It is hardly visible after a month. Here, the stem cells are produced deep within the lower layers of the skin and rise toward the surface as it is shed or injured, changing into the adult skin cell form as they go.

Skeletal muscle, like our biceps or calf and thigh muscles, have "satellite" stem cells nestled in between the adult muscle layers, waiting to be called. Intense muscle exercise alerts the satellite cells to become active and change to the adult form, bulking up and expanding the muscle. (That's why you need to leave a day or more between gym sessions, to allow the satellite cells to do their job.) All adult stem cells can continuously copy themselves, like the pluripotent ones, and then change to a specific organ type when they get the call. Some can become only one kind of adult cell while others have a wider range. Skeletal muscle stem cells can only become muscle, but blood stem cells can turn into red blood cells (which carry oxygen), platelets (which control blood clotting), and the many kinds of white cells of the immune system.

Fetal stem cells (either those we get naturally from the developing baby during pregnancy or ones that have been purified from samples of amniotic fluid) can have more potential to generate tissue than adult stem cells because of their greater youth and versatility. Stem cells gathered from the umbilical cord when a baby is born also have this greater dynamism. Parents are now paying to put their baby's cord blood samples into cell banks, storing them against the day when their child might need them for some vital repair.[5]

Our body's ability to repair is impressive, but it is outclassed by other members of the animal kingdom. Reptiles can shed and regrow skin or tails. The axolotl is a kind of salamander that has an amazing capacity for regeneration. It can regrow its tail, heart, or spinal cord and can completely regenerate an entire limb when it has been lost. This is fortunate, as axolotls have a peculiarly violent lifestyle, where chewing off each other's limbs is a daily sport.

Sadly, in the human heart (or the hearts of any mammal) we have found a very low rate of new cardiomyocyte creation. The bomb-test carbon dating evidence in chapter 3 showed us that turnover of the heart muscle was only 1 percent a year at best. When the heart is injured the amount of regeneration increases, but overall, the main response to damage is to produce a scar. Very young mammals can show regeneration of new heart muscle in the first few days after birth, because the cardiomyocytes are dividing efficiently, but this drops very rapidly after a week. Even human babies can have this ability. One study describes a newborn developing massive cardiac damage after a heart attack. Astonishingly, within weeks there was full recovery of the heart, and the baby went on to have a typical development.[6] This is strikingly different from what happens in an adult, when cardiomyocyte division drops to a low level.

If we can understand this tiny amount of natural repair that remains in the adult heart, could we perhaps stimulate it? Could we bring ourselves back to the newborn state? Scientists first needed to decide whether the repair that we do see in adults is caused by multiplication of existing cardiomyocytes or whether there is a true cardiac stem cell, like skeletal muscle. This was not as easy as it looked. The search for the elusive cardiac stem cell has been long and acrimonious, with bitter rivalries, accusations of fraud, careers ruined, and millions of dollars of research money spent in an exhaustive and exhausting quest. The dust is still settling, but we are close to concluding that any stem cell population in the heart may make new blood vessel cells, but only a handful of cardiomyocytes can be produced this way. The heart's ability to make new muscle remains stubbornly low in adults because it depends on a very small amount of cardiomyocyte proliferation. And as we shall see later, even efforts to boost this division of cardiomyocytes to increase cell numbers is fraught with risk.

MIX AND MATCH?

So, if the heart doesn't have its own stem cells, could it borrow them from other tissues? This was our next idea, to use the very active stem cells that we knew could repair blood or muscle and persuade them to become cardiomyocytes. Cardiac surgeons were the drivers here, the fighter pilots of the surgical world, eager to find something new to help their patients. They had been trying a range of imaginative surgical methods, one of which was to loosen a muscle from the back and wrap it around the heart to supply extra strength. Jump leads from the heart to the back muscle made them beat together. A leap of imagination brought the idea of taking satellite stem cells from a similar muscle, growing them in a dish, and injecting them back into the patient's heart. Because they were the patient's own cells, there was little risk of immune rejection.

At first the results looked good: there was an improvement in heart function. But then nearly half of the first small group of patients showed disturbances of heart rhythm. Doctors could see that the implanted cells had remained stubbornly as leg muscle and had not, as hoped, become cardiomyocytes. More clinical trials were done, this time using implanted defibrillators to shock the hearts out of any dangerous rhythms. But in the end the risks were too great, and that line of investigation was abandoned.[7] Bone marrow cells have gone down the same pathway, again with trials started by impatient clinicians excited by this stem cell population and reassured by its long safety record. A few thousand cardiac patients have had bone marrow extracted from their hip, treated in a laboratory to concentrate the cells, and reinjected into their hearts. Again, the use of a patient's own cells prevents rejection. The procedure is safe, with few reports of any side effects or rhythm disturbances in the heart. And there are tantalizing signs of it being beneficial.[8]

The problem is one of patient numbers, as we saw with the genetics trials in chapter 4, "Big Data—Many Hearts That Beat as One." If we have an exciting but untested new type of treatment, we must start small to protect patient safety. Individual small bone-marrow stem-cell trials, with 30 or 40 patients, sometimes showed a positive effect on heart performance and sometimes did not. But even for the drugs that are proven as valuable (and now routinely used for heart failure) the effect is not clear unless thousands of patients are tested in a trial. Our workaround for this problem is to group all the small trials together and perform an analysis on the whole set of patient information—a "meta-analysis." When we do a meta-analysis for the bone marrow heart trials, then the overall result is positive, but only by a small amount.[9] Your normal ejection fraction (the amount of blood expelled from the heart with each beat) is 60 percent and heart failure might bring it down to 30 percent or lower. The benefit given by injection of bone marrow cells is about 3 percent. It's a benefit, but it is far short of a cure. And once again, it's the blood vessel cells that are produced and not the essential muscle cells of the heart.

The scientists took a closer look at the bone marrow cells they had extracted, to try to understand what was going on. Cells extracted from some people seemed more sluggish than normal and they didn't multiply rapidly in the laboratory dish. Bad stem cells came from patients who were older and male;

who had high cholesterol, diabetes, or high blood pressure; and who smoked and didn't exercise.[10] If you think these conditions seem familiar, you are right. These are the risk factors for heart disease itself! Perhaps we had been looking at this the wrong way—was it possible that the effect of all these risk factors was to dampen some natural repair function of the bone marrow? We had been giving people back their own cells to prevent immune rejection, but in fact, what we really needed was bone marrow from someone different, someone without these risk factors. I call this the "blood of virgins" hypothesis, harking back to the old folk tales telling how to keep eternal youth and health.

THE THERAPEUTIC STEM CELL

Into the picture now came a new cell type, slightly mysterious, but with new and important properties. These are the mesenchymal stem cells, or MSCs.[11] They calm down the immune system, giving themselves a fighting chance to survive longer in a different body. Now we have the prospect of person-to-person transplant, choosing cells from healthy donors. MSCs are a small part of those stem cells we found in bone marrow, but they can also be extracted from amniotic fluid, cord blood, or fat. It's quite attractive to think that fat, which caused a lot of the trouble in the first place, could be a source of healing stem cells for heart disease.

By themselves, the MSCs can turn naturally into bone, cartilage, or fat cells. As with the bone marrow and satellite cells, hopes that MSCs will turn into cardiomyocytes have been largely dashed. However, their healing effect seems much more significant than simply their ability to turn into other cell types. MSCs have brought us a whole new idea about how stem cells can work—not only by changing into other cell types but also by releasing a flood of therapeutic factors that protect and restore damaged tissue. This is just one part of the whole conversation that is continually happening between cells in the body. A protein called FSTL1 can be released from special cells that form a single thin layer of the surface of the heart and can move through the tissue to heal and regenerate the muscle threatened by a heart attack.[12] But, as we saw earlier, the transfer of messages is much more complex and sophisticated than just the release of individual molecules one at a time. MSC have harnessed the power of using vesicles or exosomes, little packets of protein and RNA, to multiply their messages.

MSCs are certainly exosome factories, and this is important for their clinical effect on heart function. I have described how heart attack damage is amplified by the immune system, with low-level inflammation continuing to create injury even after the scar on the heart has healed. The power of MSC-exosomes to damp down the immune system works here too, soothing the inflammatory irritation. Making new blood vessels is also a critical part of the MSC-exosome effect. When a blood vessel blockage has caused a heart attack, one of the main clinical treatments is to graft large new blood vessels onto the outside of the heart to produce a bypass for the blood flow. The MSC-exosomes stimulate growth of networks of new blood vessels deep within heart walls, which improves in the same way the flow of blood to the tissue.

How far have we traveled toward a treatment with MSCs or their exosomes? Many clinical trials with MSCs have been run for different diseases and they appear to be safe, which is a good start. (Exosomes would have a big advantage as they can be treated like a drug, with no need for the complex procedures of cell implantation. But this is a newer idea and therefore not many trials have been done.) When we ask whether MSCs work for heart disease, the answer so far is rather like the one for bone marrow cells—maybe. Small benefits are seen and not in every trial, but on average it's a modest positive answer. But again, there is no evidence for creation of new cardiomyocytes in any useful number. It seems the natural mechanisms for repair of heart muscle are completely overwhelmed by the shock of the devastating damage caused by a heart attack.

So: where can we get these new cardiomyocytes?

TURNING BACK THE CLOCK ON THE HEART

I can't tell you how exciting it is for me to look down a microscope and see beating heart cells made in my own lab. To know that only a few weeks before they had been ordinary skin cells but, using the power of incredible advances in knowledge, we had literally created new heart muscle in a dish. In the early days, only the faintest flicker of life showed we had been successful. Within a few years, whole dishes of cells were pulsing in synchrony. A few years more and now we don't need microscopes—slabs of throbbing muscle as big as a thumb, and then a hand, are straining the posts where they are tethered. The science behind this is extraordinary.

It started with one of the central laws of biology being overturned. Professor Shinya Yamanaka, with only his small team of PhD students at Japan's Kyoto University, did something we all thought was impossible. They showed that the path from stem cells to adult cells was not irreversible but could be retraced: they created pluripotent stem cells from adult human skin cells![13] These behaved exactly like the precious tiny ball of stem cells in the early embryo. They could divide and multiply indefinitely, creating huge numbers of daughter cells. Then, when the conditions were switched, they could make every cell in the adult human body. Four years after the first paper came out, Shinya Yamanaka was awarded the Nobel Prize. This discovery not only gave an amazing insight into the beauty of biological processes, but it also had an immediate galvanizing effect on regenerative medicine.

Yamanaka made this huge advance by studying embryonic stem cells and pinning down four key factors that made them special. He then used harmless viruses to get them into the skin cells, which took on the characteristics of the stem cells—their clock had been reset back to the beginning of life. They not only matched the embryonic stem cells but surpassed them in several ways. First, they took away all the ethical problems and doubts associated with using something from an embryo. Many countries had banned the use of embryonic stem cells because they came from destruction of the blastocyst, the 10-day old bundle of cells that would go on to form an embryo and then a baby. Although they were mainly made from blastocysts that were surplus after in vitro fertilization (IVF) and about to be destroyed (with the consent of the parents), people felt that destroying the blastocysts to harvest cells had taken away the potential for a life to be created. The United States did not ban embryonic stem cells but prohibited the use of public funding to support their development for therapies. Successive presidents have alternately lifted and reinstated this ban. The UK allows their use, but under stringent legislation.

Now that we have Yamanaka's new "induced" pluripotent cells, the ethical constraints have lifted. Governments have mobilized huge resources, partly because of their strong desire to move away from the contentious embryonic stem cells, and scientists have been able to make rapid advances in their production and use. Not only skin cells can be used as the starting point, but also cells harvested from blood or even urine. We have designed newer and safer ways to get the four special factors into the cells. We have perfected methods for coaxing the induced pluripotent cells to become kidney, brain, liver, and heart and many other tissues.

The huge driver for induced pluripotent stem cells is a shining vision of spare organs, matched to you, made from your own spare cells. Embryonic stem cells would have become a distinct individual if they had been left to develop. So, like normal heart transplants today, implanting any heart tissue made from embryonic stem cells would need lifelong immunosuppressive drugs. These come with a price—an increased susceptibility to infection, for example. But induced pluripotent stem cells made from one person are immunologically matched to that person, and therefore they should not be rejected if implanted back. This is an amazing breakthrough and a complete game-changer for regenerative medicine. However, as we shall see later, the devil is in the details.

OWN-BRAND OR DESIGNER LABEL?

My excitement about making cardiomyocytes from induced pluripotent stem cells (iPSCs) was shared by scientists all over the world. We had two challenges—to make them as good as possible and in large numbers. Remember, our mission is to replace up to a billion cells in every damaged heart. We were taking our cues from the biology of the developing embryo and trying to mimic the changes in hormones and protein growth factors that shape the emergence of the beating heart, one of the earliest organs to appear. Every parent remembers that tiny flutter of the heartbeat on the first scan—the baby at six weeks is no larger than a grain of rice.

First, we used mixtures of the growth factors that the developing embryo would have experienced. As we understood more about the mechanism that drives the changes, we selected precisely targeted drug-like compounds to switch on and off the signals at the right times. Within a few years we went from hit-and-miss methods to developing highly reliable systems that will work in any competent laboratory. iPSC-cardiomyocyte production is now consistent and reproducible, which makes us confident that we have a true therapeutic product and not just a scientific curiosity.[14] Companies sell them and ship them around the world.[15] If you can order custom-made cardiomyocytes out of a catalogue you know this is now reality.

But how do the iPSC-cardiomyocytes measure up to the adult cardiomyocytes we know and love? We brought to the problem all our tools of electrical measurement, optical mapping, advanced microscopy, and big data analysis of their fluctuating biochemistry. They are real cardiomyocytes, but they don't have an exact equivalence in natural development. They are most like very young cardiomyocytes, as you might expect from something that has just come from an embryonic-like cell. This has pluses and minuses. The iPSC-cardiomyocytes are much tougher than the adult ones: when we culture the adult cells in a Petri dish, they last for a few days only. Compare this with the iPSC-cardiomyocytes, where we have had clusters beating for over a year. We had a little birthday party for them—scientists know how to have fun! And you can send them through the post—I received some from Japan that arrived floating in a tube and started beating as soon as we warmed them up.

On the downside, the iPSC-cardiomyocytes are less powerful than the adult cells in generating force. The iPSC-
cardiomyocytes are thinner and with less muscle protein, and
lack the deep-diving t-tubules that bring the electrical impulse
down into the cell. This gives them a slower contraction and
relaxation and less force from the muscle. iPSC-cardiomyocytes
also use simpler ways to generate energy, sucking pure glucose from the culture solution instead of generating power
from the mitochondria inside. The reason for this mechanism
in young cells is to protect them in the low-oxygen environment of the womb. When we think about implanting them
into cardiac muscle after a heart attack, this immature energy
metabolism may protect the iPSC-cardiomyocytes from the
lack of oxygen they will encounter there.

When we set about finding ways to improve the iPSC-
cardiomyocytes, we immediately realized we had to give
them back the 3D working environment of the heart. We had
known for many years that depriving the adult cardiomyocytes
of their neighbors was very damaging—alone in a dish they
deteriorated and died. iPSC-cardiomyocytes needed a community to grow and develop. Luckily these baby cells, unlike
the adults, are very good at finding friends. The trick we use
is to envelop many iPSC-cardiomyocytes in a soft gel made
of fibrin, the protein you get in a blood clot. Then we add
thrombin, the protein that shrinks down the clot and makes
it firmer. The cells are forced closer together and they naturally reach out to find and link to their neighbors. At first this
sheet of cells is completely still. We know that each individual
cell continues to beat, but they are all out of synchrony with
each other. Then after a few days the sheet begins to quiver as
little clusters of cells set up neighborhoods. After a week they
have all found each other and the gel begins to beat as a single

sheet. Lo and behold, we have made our own beating cardiac muscle—engineered heart tissue![16]

Now we need to give the encased iPSC-cardiomyocytes some exercise. The heart gives a cardiomyocyte constant mechanical stimulation—it's the lack of that stimulation that accelerates their deterioration when they are isolated in a dish. So we add to the gel some bendy posts. When the engineered heart tissue contracts with each beat, it pulls the posts inward. When it relaxes between beats, the posts spring back and stretch the tissue. Like weights in a gym, pulling against the posts makes the muscle tissue grow stronger over days and weeks. With vigorous exercise and extra pacing to increase the rate of beating, the engineered heart tissue is getting close to exerting adult forces. When we test the engineered heart patches in animal models of heart disease, we can see that the patches stimulate real improvements in heart function.

We can make these engineered heart tissues as long or as wide as we want: thumb-sized patches like a Band-Aid for the heart, to place over spots of damage, or large hand-sized ones to wrap over the place where the scar has formed from the heart attack. It's harder to make the heart tissues we have created any thicker, as oxygen will have problems getting into the middle. As we struggle to make the iPSC-cardiomyocytes more like adult cells, we risk losing their unique ability to survive in a low-oxygen environment. It's a difficult balance. But we can get half a billion iPSC-cardiomyocytes into the palm-sized patch—well on the way to replacing the number of adult cardiomyocytes lost in a heart attack. And though it costs us around $15,000 to make one (which is a stretch for a small lab), the price is low when compared to the latest cancer drugs or a partial artificial heart assist device.

We now have patches of engineered heart tissue, which can be made from the patient's skin cells and expanded to a clinically useful size, and have been tested and shown to work in animals. So what's the problem? Why aren't they used in hospitals right now? This is where we hit the so-called "Valley of Death"—the huge gap between a successful experiment and translation to a real-world product. This gap is hard enough to bridge for introducing new drugs, where we have decades of successful experience. For novel therapies such as engineered heart tissue we have a heap of new problems that extend from the biology itself, through clinical logistics, to the paradox at the center of the economics of regenerative medicine.

PROBLEM #1: THE BLOOD OF VIRGINS AGAIN?

Scientists keep learning the hard way that it's very difficult to beat the immune system. When we transplant hearts, we also give the patients receiving them an immense burden of lifelong immunosuppression to keep these hearts from being rejected. It's been more than 50 years since the first heart transplant and still these immunosuppressant drugs, with their toll of side effects and increased infections, are the best we have. The promise of new iPSC-cardiomyocytes that are immune-matched to the patient is the Holy Grail of regenerative medicine. But it's not as simple as that, as we are finding out.

First, as with the bone marrow cells and the "blood of virgins" hypothesis, the starting cells taken from the patients to make iPSC may be substandard in some way and someone else's would be better. I have shown you how there are hidden heart mutations that are common and have damaging effects, not seen until some other stress sets them off. People who

have heart disease may well be more likely to have these mutations, and the defective genes will also be there in the skin
cells that are reprogrammed to stem cells. Could the effects
of these mutations be seen in the iPSC-cardiomyocytes? We
know they can, and this has produced a whole new branch of
science—called "disease in a dish." As you will see later, this
has been a big win, almost eclipsing the promise of cardiac
repair. But it's a real barrier if some patients can't have this life-
saving treatment because their heart defects carry through and
the engineered heart tissue is dysfunctional.

Even with the best iPSC, we hit the immune system barrier
again. Is it dangerous to implant immune-matched cells that
can grow and expand in number indefinitely, as the iPSCs do?
This is basically the description of a cancer. In fact there is a
specific, and rather gruesome, cancer associated with pluripotent stem cells. Called a teratoma, it is a random mix of hair,
teeth, bone, and many other tissues. When the iPSCs have
successfully become heart (or other tissues) they no longer
make a teratoma and there is no danger. But how many con-
taminating, unchanged, iPSCs left in the iPSC-cardiomyocyte
preparation would be dangerous—a thousand, a hundred,
one? Until we have data on implants in millions of people,
we will not know the true danger of this tiny but possibly
significant risk.

On the other hand, the skin (or other) cells we take from
the patient start off as immune-matched, but do they stay that
way? We keep them in culture for months, treat them with
factors to make them into completely different, induced pluripotent cells, and then we expose them to many chemicals
to make them into iPSC-cardiomyocytes. There are lots of
chances for them to change biologically through this process.
We find that they can start to acquire fetal proteins on their

surface as we turn back their biological clock to make them stem cells. Some of these are not usually around when the body is making its inventory of "self" vs. "non-self." Cells with these proteins are recognized by the adult body as foreign and the immune system destroys them. We can do some tests on immune cells in a test tube, but we can't really be sure that the iPSC-cardiomyocytes are a true match until we implant the engineered heart tissue into the first patient.

PROBLEM #2: SORRY, YOUR HEART IS IN THE MAIL

The next dose of cold-water reality is in the logistics of the iPSC-cardiomyocyte therapy. This is truly personalized medicine because it is absolutely centered on the individual patient. Let's think about Martin, as he is being rushed into the emergency room, unconscious and in the throes of a serious myocardial infarction. Do we want to give him an iPSC-cardiomyocyte patch, knowing as we do that a large part of his heart is dying even as he is lifted on to the table? We don't yet know anything about him. Is he diabetic? Does he have an arrhythmia-causing mutation? He might not even know these things himself. We don't know how well he will do anyway with our state-of-the-art cardiac rescue protocols. He might be one of the lucky ones who came in early, got the most rapid stent or bypass operation and the best drugs, and had most of his heart muscle snatched back from the brink of destruction. These are all good reasons not to rush in with our new patch. But most of all, we don't have time.

Biology can't be hurried; it takes how long it takes. We can turn skin cells into iPSCs in a matter of weeks. Then we must allow these to grow and divide into half a billion cells before we go further (two or three months at best), because the iPSCs grow well but this slows right down when they become iPSC-cardiomyocytes. Making the half billion iPSCs into cardiomyocytes is another three weeks, and then two months of exercising them to get the best, strongest engineered heart tissue. Six months at least are required to get a personalized therapy to a patient: clearly this is never going to be an emergency treatment. Probably, we are aiming for the patient who does not do well, who starts the tragic slide into heart failure months or years later, despite all the best efforts of the doctors. If we pursue this direction, we will be putting the engineered heart tissue over a deeply scarred and damaged heart. That's a tough ask for any therapy.

But if biology is unhurried there is something worse to factor in, and that is government regulation: if the complexity of the science doesn't get you then the paperwork will! We must make sure every cell, solution, piece of equipment, lab construction, and protocol are certified as fit-for-purpose and compatible with future clinical use of the product. iPSC-cardiomyocytes must be made under Good Manufacturing Practice rules, a highly stringent set of criteria that will test each cell line for (among other things) purity, sterility, absence of unchanged iPSCs, and a complete shopping list of ideal cardiomyocyte features. To fulfill this for one product is a significant barrier to jump. To do so individually for every single patient is a huge mountain to cross over.

Our ideal solution would be a universal donor line, like the O blood group for blood transfusions. Then we could find the ideal iPSC line or lines, with no known mutations; work hard to get the most authentic iPSC-cardiomyocytes; exercise our engineered heart tissue into peak fitness; make a customer-friendly range of different-sized patches and store them safely for off-the-shelf use. Plus: we would jump all the regulatory hurdles while doing all of this. This is where the science is heading now. Scientists can now perform very precise editing of genes (I will talk later about the revolutionary CRISPR/Cas9 technology) and so can pick off the surface proteins on the iPSCs that trigger an immune response. This is not a simple task—as mentioned, the immune system is not easily fooled—but we are steadily heading in the right direction. However, if there is one more barrier that's bigger than science and paperwork, it's figuring out how to make a profit.

PROBLEM #3: SHOW ME THE MONEY

Ironically, the biggest problem for investment into regenerative medicine is that it might work. Big Pharma thrives financially on something that will keep you alive but only if you keep taking it (and paying). The longer you must take their drug the more profit there is, which is why cardiac drugs like statins and blood pressure medications are their absolute favorite. A therapy that can be given once and completely cure you will have to be very expensive indeed to make it worthwhile to develop. That's why the first gene therapy product for spinal muscular atrophy in infants, Zolgensma, started on the market at $2 million per dose. Desperately ill children, who would otherwise die within a few years, were dramatically rescued by a single treatment. This startling price was justified as cost effective to give, potentially, a whole disease-free life.

Cardiac patients will be older, but still have the possibility for an extra 30 or 40 years of life if the engineered heart tissue or other stem cell implants work well. How much should a company charge for this? What will the different health care systems be prepared to reimburse? Will they lose out on the money they might gain from supporting a cardiac patient for life on the five or six different drugs they are now taking? A business model where the user supplies their own cells also raises another set of questions for the companies.

Big Pharma has been slow to test the water of regenerative medicine both because of the uncertainty about whether it will work and the concerns about the surrounding business model. For the cardiac area this is even more acute, because here even the straightforward drug trials are becoming excessively expensive. Without the power of their investment and infrastructure the regenerative medicine trials for the heart will never get off the ground. A recent clinical trial of the bone marrow therapy was meant to recruit enough patients to give a definitive answer to whether the small benefits of the treatment translated into lives saved. But this trial struggled to complete its mission with only academic support and EU funding. Meanwhile, governments fight to stop the growth of maverick clinics that use the buzzword "stem cells" to sell unproven and unregulated treatments to the public.

Things are starting to change: first tentative steps are being made. The success of gene therapy, for spinal muscle atrophy, and the genetic modification of immune cells to fight cancer are showing that investment in novel therapies can pay off. Cell therapies are attracting venture capital and Big Pharma companies are setting up collaborations with academic laboratories. Governments including Japan are fast-tracking novel therapies and many countries have special exemptions for small-scale studies on compassionate grounds. We hope that one big win will tempt the hesitant companies to take the plunge!

WHEN WILL THE FUTURE BE HERE?

A strong message of this book is that cardiologists have done very well by making friends. They have reached out to oncologists, rheumatologists, psychologists, and virologists. They have gone beyond biology and co-opted epidemiologists, sociologists, engineers, and data scientists to their cause. Cutting-edge developments in stem cell biology, cancer drug development, device miniaturization, wearable technologies, and submicroscopical imaging have all been channeled toward cracking the problem of heart disease. So how well are we doing? And what will be the next big breakthrough?

THE DRUGS DO WORK

Sometimes things are better than the statistics make them look. When we see the figures for heart attacks or heart failure, we are often looking over a 5- or 10-year period. But not all of those years are equal: you might expect that the most recent years are possibly better than the earliest. There are hints that this is so when we look at the patients we have right now in hospitals and the community. Often, when we come to set up new clinical trials, it can be hard to get enough of the patients we expect. Clinicians try to assess the number of patients they will need by looking at published data on the prevalence of the condition. They may select ones that have disease in either a moderate or severe form, depending on whether they are trying to measure symptom improvement or mortality rates. They work out what sort of improvement would be clinically meaningful—a 10 percent increase in cardiac output or a 20 percent improvement in survival, for example. Then they calculate how many patients it would take to be able to show statistically that this has been achieved. All this planning tells us whether the trial has a good chance of working.

However, we are increasingly getting the numbers wrong. A recent trial to study regeneration with bone marrow cells aimed to recruit 3,000 patients.[1] The researchers were looking for people who had had a heart attack and whose left ventricular ejection had fallen from the normal 60 percent down to 45 percent. Mortality rates in these patients would usually be quite significant and they were hoping to be able to detect a 25 percent rescue from death with the cell treatment. But even with 12 large hospitals from all over Europe participating, only 375 patients with disease this severe could be recruited over the course of several years. Even in those patients, the sickest the researchers could find, the death rates were small: six people died in the treated group over two years compared to seven in the control. The trial was a technical failure, but the underlying snapshot of the improving recovery of heart attack patients was a pleasant surprise. In fact, it is often the case that the placebo group members in clinical trials do very well. Because it is important that the control group is given the best possible current treatment, to give a fair comparison to the new therapy, every care is made to supply the full range of cardiac drugs. This trial and others have shown that when the cardiac drugs are given in a supervised way to ensure the optimum therapy, they really do work.

A huge question is whether these drugs are curing the disease or just keeping it at bay. Withdrawing the drugs would tell us this, but until recently no one had dared to do it. However, some patients, especially the younger ones who had responded well to treatment, began to ask whether they needed to be on drugs for life. It may be that they were finding side effects unpleasant, or they were women who wanted to start a family and didn't want to be on drugs while pregnant. So, a study cautiously withdrew the drugs one by one over six months from patients who had responded well, had good heart function, and were stable.[2] Out of 51 patients, 20 showed signs of relapsing but recovered when the drugs were given again. The other 31 were apparently well and able to remain so without the drugs, although so far this has only been continued for a short time. This is good news and hints that reversal of the disease could be possible.

Even very sick people may be able to pull back to normal heart function, given the right circumstances. Patients who are in end-stage heart failure, with very low cardiac output and massively enlarged hearts, are now often implanted with a mechanical left ventricular assist device or LVAD. The LVAD supports the circulation of the blood and, with time, clinicians have observed that the heart seems to show signs of recovery. The size of the heart decreases and, when the LVAD is temporarily turned off or down, the heart takes over pumping the blood. Occasionally, it has been possible to turn the LVAD off completely or even take it out again. A group in Kentucky has now tried to do this systematically by carefully controlling the LVAD power to get the best reduction of heart size.[3] Crucially, they also gave patients on LVADs large doses of the beneficial cardiac drugs. Without the LVAD, these would have been too strong for the patients to tolerate, but the LVAD allows them to do that. Twenty patients have now had their LVAD turned off or removed and have been stable with normal heart function for up to two years. This is an incredible recovery, from close to death to a normal life.

New drugs are also on the horizon, attacking the problem from many different angles. Simply working on the cancer drugs to reduce their toxicity for the heart will prevent many cases of chemotherapy-induced heart failure. For cardiac drugs, most of the successful ones in use now target the body's hormonal response to low cardiac output—the beta-blockers, ACE inhibitors, and diuretics. These are being improved and refined, with further small but important benefits seen. Inflammation blockers, which also target the body's attack on itself, are a new class of drugs to protect the heart. In the extreme danger of the immune storm after COVID-19 infection, drugs such as dexamethasone saved lives in the intensive care units. For the more prolonged, low-level inflammation after a heart attack, recent clinical trials have succeeded where others had failed by carefully targeting very specific molecular pathways.[4]

Moving past the protective drugs to ones that give immediate positive benefit and improve symptoms has been more challenging. Drugs that stimulate heart function and increase blood flow have been stymied by rhythm disturbance side effects. We know that the main stimulatory pathways used by the body, like those activated by adrenaline, can often cause arrhythmias themselves, which is why we inhibit them with beta-blockers. So we need to avoid those pathways when we design new drugs. Some successful agents have targeted the mechanics of muscle contraction directly, but it's a fine line between stimulating contraction and interfering with proper relaxation of the heart. Often, we know a great deal about the pathways and what we need to do but cannot find a drug to do it. Most drugs consist of small molecules or antibodies, and they work well when there is a specific receptor on the cell surface to which they can bind. But in other cases, we want to disrupt (or help) interactions between large proteins inside the cell. The smaller molecules often cannot do this, because of their size or inability to penetrate the cell membrane. If we could target the genes themselves, we could change their protein products directly. This is where we have turned to the newer sciences of gene therapy and, even more recently, gene editing.

Although the heart poses special challenges for gene modification, I could not write a science book now without talking about these new techniques. They expand the horizons for medical science immeasurably. They give hope that even the inborn errors in genes, which cause devastating and incurable conditions, could be prevented or reversed. No longer would parents have to make the impossible decision to abort a fetus or to care for a severely disabled child for many years. Patients with debilitating conditions such as hemophilia would be able to live a normal active life. The science is complex, but the potential rewards are immense.

THE GENE GENIE

Gene therapy was the first venture into this virgin field. This involves introducing genetic material such as DNA and RNA into the body so that it can function like a factory to produce new proteins. The first COVID-19 vaccines were produced so quickly because they used this technology, harnessing body systems to make the coronavirus spike protein from RNA, instead of requiring the laborious process of growing and purifying the viruses themselves. But it was a long journey from a precarious start to get to that point.

Gene therapy started in earnest in the 1990s with a number of successes, but was suddenly halted by one disastrous event. Getting the DNA or RNA into the body had been the first hurdle, and at last the scientists had learned how to do this, by copying our natural enemies—the viruses. Viruses are tiny strings of DNA or RNA, often contained within a spiky coat that allows them to latch on to cells. (The ominous ball and spike shape of the coronavirus is now etched into the imagination of the world.) Researchers can take common cold or other everyday viruses and make them safe by removing the genes that allowed them to cause disease or multiply. These are then replaced with a gene that would encode for a desired protein. The new modified virus DNA or RNA is injected like a vaccine and harnesses the cells' own machinery to begin immediately making many copies of the protein.

The first successes were for children who were born deficient in proteins related to the immune system. On September 14, 1990, in Maryland, the first gene therapy treatment was administered to a four-year old girl called Ashanthi DeSilva who had severe combined immune-system deficiency (SCID) due to adenosine deaminase (ADA) deficiency. Children with this condition are dangerously vulnerable to infections: often they need to live in a sterile room, or bubble, for their entire lives. A modified virus was used to insert the missing gene into the DNA of Ashanthi's white blood cells. An extra safety aspect was that the blood cells could be treated outside the body, to avoid injecting the virus directly, and then cells could be reinjected. This was successful, and 18 more children over the next decade were also cured in this way.[5] Other children with different forms of SCID were treated, with equally promising results.

In the first flush of excitement, scientists and clinicians were keen to roll this out as far as possible. In September 1999 Jesse Gelsinger, a teenager with ornithine decarboxylase (OTC) deficiency, entered a clinical trial at the University of Pennsylvania. OTC deficiency results in toxic ammonia production from breakdown of protein, but the milder form that Jesse had could be treated by drugs and a low protein diet. He was keen to be treated with this new therapy, saying, "What's the worst that can happen to me? I die and it's for the babies." On September 13, 1999, Gelsinger was injected with a viral vector carrying a corrected OTC gene. But the dose was too high and he suffered a massive immune response to the virus itself, leading to multiple organ failure and brain death. The researchers involved were found to be guilty of ignoring or concealing previous results that might have predicted this response. This was a shock for the field and slowed down research on gene therapy for many years. A further blow was the development of leukemia in some SCID children, where the virus had integrated into their DNA at a sensitive spot.

The gene therapy field has only slowly recovered from these initial setbacks. Gradually over the years the viruses have been refined for maximum safety and clinical trials have begun very cautiously. Positive results have accumulated, and confidence has grown. Cells infected outside the body and returned to the patient have produced successful results. One of the most exciting new cancer treatments, CAR-T cell therapy, involves taking a patient's white blood cells and infecting them with a virus that will stimulate them to attack the cancerous cells. Cancers like acute lymphoblastic leukemia in children, and others that have a high rate of relapse after drug treatment, are starting to benefit from the CAR-T cells.

Defects in proteins that circulate in the blood have also been a more tractable target for gene therapy. Blood-clotting proteins, which are missing in hemophilia patients, have been restored using a virus called AAV (adeno-associated virus). This respiratory virus is quite common in the population but produces little or no obvious illness. A single dose of the modified AAV virus, engineered to produce the missing protein that controls blood clotting, has been able to restore protein levels for years in animals.[6] Patients who were previously dependent on daily injection of the clotting factor, and in danger of hospitalization with episodes of bleeding, extreme joint pain, and fever, could now live a normal life.[7] Going back to the ADA-SCID patients like four-year-old Ashanthi DeSilva, who started the gene therapy journey, the gene therapy vector for ADA (now called Strimvelis) is the first of an increasing number to be given regulatory approval for use as a medicinal product. The portfolio of new gene therapy agents is growing monthly.

BATTLES AT THE CUTTING EDGE OF GENE THERAPY

Developing gene therapy for solid organs like the heart is proving difficult, as my personal story of struggle and failure will show. Once again, we learned the hard way that the heart defends itself from all of scientists' best efforts to change its makeup. We wanted to improve the contraction of the failing heart, in a way that would not have the dangerous effects on heart rhythm that had hindered all the previous new drugs. I started my gene therapy journey in 2009, when evidence emerged for a safe new heart-targeted virus. Our aim was to restore levels of SERCA2a, one of the proteins that controls calcium storage inside the cell. We knew that this was lost in the failing heart, and our experiments on human cardiomyocytes showed that restoring SERCA2a using the new virus strengthened the contraction of the cell.[8] Not only that, but the relaxation was also faster, and the cardiomyocytes were more resistant to arrhythmia. This would be the ideal combination of effects to bring the failing heart back to normal.[9] The regulatory-required experiments in animals were positive, with an excellent safety profile and beneficial effects on heart function.

We began our negotiations with the regulatory bodies that oversee new therapies with great hope. These regulatory bodies were now being faced more and more with what are called "advanced therapies": not the customary drugs but rather, cells or viruses. These are different in many ways from the conventional small-molecule drugs. For example, they are irreversible. In a normal clinical trial, if the drug causes side effects you stop administering it. But the virus will have delivered its load of DNA, and this may stay in the body for many years. It will keep producing the new protein as long as the DNA remains, with no way to remove it. The body will also produce antibodies to the spiky virus coat. So, if you give a patient a low dose of virus you cannot later give a higher dose, because they will now have antibodies that neutralize the virus. Patients in the early trials, which used this low dose, may not be able to benefit from the final therapy because they still have these antibodies. All these complications mean that the whole ethical balance of safety, harm, and benefit must be completely reconsidered.

Because of these complexities, the regulatory hurdles were enormous: it took around seven years to obtain approval for

clinical use of the therapy in the UK, and some of the consequences were unexpected. At one point, it was necessary to
comply with the regulations intended for genetically modified
crops, since the infection of patients with the virus and their
journey home from the hospital was classed as a "deliberate
release of a genetically modified organism." We had to publish
the map coordinates of the hospital and put this out to public
consultation!

The first trial using the virus to carry SERCA2a to the heart,
called CUPID, was published in 2009, and showed safety—
plus a hint of benefit—for the heart in nine patients.[10] Later,
more patients were added to determine the best dose. This was
a great boost, and we were able to join in a follow-up of this
trial with a larger study of 250 patients, called CUPID 2.[11]
We also set up our own small study on patients who had had
left ventricular assist devices implanted, since many severely
ill cardiac patients were now on these devices. We learned
many valuable lessons, mostly the hard way.[12] For example,
in the UK, 60–70 percent of patients were not eligible for the
trial simply because they already had antibodies to the natural
virus. The cold, damp winters of the UK had been an ideal
breeding ground for respiratory infections and therefore many
people had had this common virus without realizing. But the
hardest lesson, and one that caused the failure of both trials,
was the strong resistance of the human heart muscle to infection by the virus.[13] Tissue samples showed that few if any viral
particles had gotten into the cardiomyocytes. We were beaten
again by the impenetrable heart!

We have been back to the drawing board. We need higher
doses of virus, and we must redesign it to have a different
outer coating from the natural one, so that preexisting antibodies do not block it. Maybe we need multiple varieties of

coating, so that we can give more than one dose of virus. We need to engineer in a "kill-switch" so we can turn off the virus if we see any danger signs. But we must carry on, because otherwise all our knowledge of the mechanisms of heart failure will be wasted.

There have also been exciting new developments in gene editing, using the CRISPR/Cas9 system to snip out single genes very precisely. This technology, called "molecular scissors," lets us repair, add, or delete genes in a way that avoids unwanted effects on the remainder of the DNA. Professors Jennifer Doudna and Emmanuelle Charpentier at UC Berkeley were awarded the Nobel Prize for their breakthrough research on CRISPR technology. This gives us hope for gene therapies, and as you will now see, CRISPR is already turbocharging the stem cell technologies.

MUSCLE FOR THE HEART

Advanced therapies, which promise not just a treatment but a cure as well, also encompass the new stem cell therapies coming into the clinics now. We have talked about drugs or devices to preserve heart function, and how they are allowing some real recovery, as well as gene therapy to stimulate cardiomyocyte contraction, and how that might boost cardiac output. But both these strategies only work on the surviving cardiomyocytes. Massive damage from a heart attack will leave too few cardiomyocytes to cope with the body's needs, whether they are working at normal or even super-normal levels. What are the very latest strategies to get back those missing cells?

First, there is the idea to stimulate that 1 percent of turnover of cardiomyocytes we saw from the carbon dating studies. Researchers have gone back to the microRNAs (miRNAs), the short sequences of genetic material that are the orchestral conductors of all the body processes. These are the tiny controllers that help the same 20,000 genes to produce an onion or a human being. Since there are fewer than 2,000 miRNAs, it is possible to create laboratory models to test them all individually. This strategy highlighted a handful of miRNAs that might have the right characteristics to stimulate cardiomyocytes to multiply, when tested on these cells in the laboratory dish. Cardiomyocytes were now able to divide and grow continuously, producing new generations of cells. The most promising miRNA made it to preclinical testing in pigs, which are often used since there are many similarities between a pig's heart and a human heart. The miRNA was targeted to the cardiomyocytes using an AAV gene therapy vector.

Initial results seemed very promising: there was a striking regrowth of cardiac muscle in the early days after the heart attack.[14] However, the AAV+miRNA combination was too powerful and long-lasting—it just kept on growing new muscle. Every time the cardiomyocytes divide, they must break down their structure and become like immature cells to separate properly. While they are in that state, they disrupt the smooth flow of electrical current across the heart. As ever more cardiomyocytes were dividing, the heart was becoming very vulnerable to disturbances of beating rhythm. One by one the pigs were developing catastrophic arrhythmias and were precipitated into sudden cardiac death. This dramatically underlines the message that the heart is built to resist regeneration because its structure depends crucially on the perfect integration of its adult cardiomyocytes. Now the challenge is to control the therapy, walking the fine line between healing the infarct scar and tipping the heart structure into chaos. Given the potential for disaster, using the miRNA therapy will need to have many safeguards built in before it can reach the stage of clinical trials. Short-term stimulation with agents such as chemically modified RNA—already being tested in clinical trials—may lead to the answer.[15]

If we cannot safely induce the heart to regenerate itself, can we add the new induced pluripotent stem cell-derived cardiomyocytes (iPSC-cardiomyocytes) and hope that they take over some of the work of contraction? These are the cells that have been produced from a patient's own skin or blood by genetic reprogramming and are therefore immune-matched to that patient. Human iPSC-cardiomyocytes have been the great success story of the cardiac stem cell field, able to be produced in the order of billions of cells, and easily formed into beating muscle patches. Surely, they can be harnessed to supply the needed cardiomyocytes?

Once again, early experiments were promising—although many cells died when they were injected directly into the heart, the ones that remained linked up with the adult cells after some weeks or months and seemed to help restore the heart function. However, when researchers progressed to larger animals and scaled-up grafts, they ran into trouble. Severe arrhythmias started to fire off almost immediately after injection of iPSC-cardiomyocytes and went on for several months.[16] Eventually the heart rhythm calmed down, but for a future therapy this would be a very dangerous period of heart disruption. We are not certain why this is, but there are several possibilities. First, we could see that the iPSC-cardiomyocytes continued to beat spontaneously after implantation and so may have been setting up a rival and interfering pacemaker system. Second, the iPSC-cardiomyocytes are an immature kind of cardiomyocyte and therefore, like the dividing cardiomyocytes after miRNA therapy, they may have disrupted the heart structure. Two months later, they become more mature and make good connections with the adult cardiomyocytes, so the heart muscle is once again a smooth structure controlled by the main heart pacemaker. Can we solve this problem and avoid this early period with immature, unconnected, beating iPSC-cardiomyocytes?

Engineered heart tissue—the beating muscle patches—seem a way forward. They can be scaled up to the size needed for a large animal or human heart and contain a billion or more iPSC-cardiomyocytes. Being in the engineered tissue and performing work will make the iPSC-cardiomyocytes more mature than lying in a dish. Laying iPSC-cardiomyocytes on the surface of the heart and attaching them by sutures allows us to position them to bridge between scarred and unaffected tissue. Results are good even in very large constructs, with improvements in the heart function.[17] Crucially, they do not seem to cause arrhythmia. However, neither do they directly couple electrically to the heart (this may in fact be why they do not cause arrhythmia). There is a huge effort to improve these patches, by adding supporting cells of other types or by making novel materials that will improve them in terms of strength, flexibility, and electrical conduction. At Imperial College London we are trying to incorporate an electronic pacemaker to couple the engineered heart tissue with the host heart. Both the tissue engineering and the mechanical devices have their flaws—maybe the way forward is with cyborg devices, a mixture of tissue and electronics or robotics to combine the best of the two.

Tentative efforts have started to bring pluripotent stem cell cardiomyocytes to the clinic. A study of six patients had a patch implanted with pre-cardiac cells from pluripotent stem cells.[18] News reports of two patients in China with injected iPSC-cardiomyocytes and one patient in Japan with a wafer-thin engineered heart tissue sheet have appeared.[19] Several other trials are starting in Germany and France. In the meantime, however, we have found another use for the iPSC-cardiomyocytes that could help us with heart disease in whole different way.

PERSONALIZED STEM CELLS—
THE CLINICAL TRIAL IN A DISH

As well as the promise of new muscle for the heart, another use of human pluripotent stem cells has taken the pharmaceutical industry by storm. This is the disease-in-a-dish model, which can test your heart with thousands of drugs simply by taking a blood or skin sample. When you change the blood or skin cell into iPSCs, they still carry the same genes. When you push the iPSCs into cardiomyocytes, liver, brain, and so on, those genes remain unchanged. If you have a gene that predisposes you to disease in that organ it will show up in the function of the iPSC-derived cells. This means we can look at your heart function without ever touching your heart.

Everything to do with iPSC sounds like science fiction, but this part of the work has been proven repeatedly. If you make iPSC from patients who have an electrical fault in their heart that gives them arrhythmias, or makes them at risk for sudden cardiac death, then the iPSC-cardiomyocytes show the same effect. Collaborating with clinicians, we obtained skin cells from a family in Spain who had a history of heart rhythm disturbances. There was the father, who had a mutation causing abnormal heart growth and arrhythmia, one son who was a carrier of the gene defect, and another son who was not. When we created iPSC-cardiomyocytes from them, and engineered heart tissue from those cardiomyocytes, we could clearly see the waves of arrhythmia coursing up and down the muscle strips.[20] The father and affected son showed these very clearly but iPSC-cardiomyocytes from the unaffected son were completely normal. This is just one of hundreds of examples where a heart defect shows up clearly in a dish of iPSC-cardiomyocytes.

Combining CRISPR gene editing with iPSCs is an incredibly powerful technology for understanding how mutations affect the heart. In the Spanish family, when we edited out the mutant gene from the iPSCs of the affected father and son, we removed the arrhythmia. Conversely, when we added the same mutation to the unaffected son's iPSC-cardiomyocytes, they now became disturbed by a very similar arrhythmia. This was strong evidence that the mutation caused the arrhythmia. But the son carrying the mutation had a less severe disease than the father, and it is often the case that one member of a family will be much more impacted than another. We think that there is a version of the "second-hit" phenomenon we saw earlier with the titin mutation, where other genes' variations can magnify or reduce the effect of the main mutation. The combination of disease in a dish and gene editing opened a way to study this effect. We could edit the mutation into iPSCs from many people, to see if other genes could make the arrhythmia worse or less bad. Finally, we can start to understand why the same mutation can be devastating in one person and hardly noticed in another.

Drug discovery has been revolutionized by this disease modeling with iPSC (figure 11.1). First, small studies were able to test drugs on iPSC-cardiomyocytes in the dish and suggest successful ones to bring back to treat the patient.[21] Even genetic mutations present from birth might be treated this

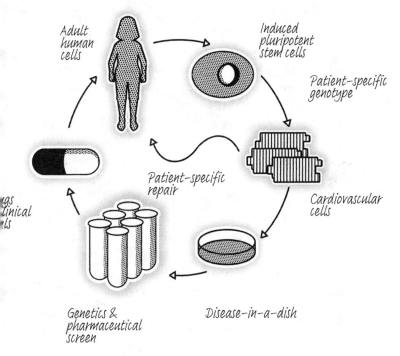

Figure 11.1
Adult human cells can be reprogrammed into induced pluripotent
stem cells, and from there turned into cardiovascular cells. These have
the potential to be generate cardiac tissue matched to the patient and
used in cardiac repair. They can also be used in disease-in-a-dish
investigations and treated with drugs to understand their safety on the
iPSC-derived cells or their ability to treat disease.

way. A team at Stanford University studied four generations of a family with a dangerous mutation in a gene called LNMA that caused a heart arrhythmia at a young age. Family members with this mutation typically had an enlarged and failing heart and the arrhythmia often led to sudden cardiac death.[22] LNMA is a protein that acts to allow cells to adapt to their environment by sensing mechanical changes. The mutation in this gene affected both cardiomyocytes and the lining of the blood vessels (endothelial cells), leading to fibrosis and scarring of the heart. Researchers took tissue samples from family members and reprogrammed them to iPSC. They then created both cardiomyocytes and endothelial cells from the iPSCs to study their function, either separately or interacting together. Comparing the range of protein expression between these cells and cells from unaffected family members, they were able to identify differences in expression of proteins. Advanced mathematical tools let them hone in on likely targets. Gene editing of the iPSC-cardiomyocytes or endothelial cells helped them confirm which changes were important in the disease process. A protein named KLF2 seemed to be the key. Analyzing structures of drugs already in clinical use, they selected ones shown by mathematical modeling to bind to KLF2. Some by chance were statins: as well as reducing cholesterol they had this additional property of modulating KLF2. Clearly, they would be a safe and beneficial drug to give to a cardiac patient anyway, making them a likely set of candidates to be tested. Testing on the iPSC endothelial cells and cardiomyocytes confirmed that lovastatin, one of the best of the candidates, reversed the damaging effects of the LNMA mutation. Finally, the best test of all, treating a small number of the patients showed that treatment with lovastatin improved their endothelial function. This is a triumphant example of the new science at its best, combining stem cell biology, mathematical sciences, and clinical medicine to have a real-world impact on a previously incurable disease.

Now, with the power of making billions of iPSC-cardiomyocytes, large-scale runs can be done where hundreds or thousands of potential new compounds could be tested at once. Drug companies can select successful compounds much more efficiently with this human-based, patient-specific cell source. The best drugs can be tested on iPSCs from many individual people, both with and without heart disease: a clinical trial in a dish! The cells can be "mixed" with other cell types in the heart to make a more realistic comparison to the in vivo setting; they also can be "aged," again to make a more relevant model to adult hearts. Animal testing can be drastically reduced: much of this work has been supported by organizations that seek to reduce animal use in science. Although the final regulatory steps for a drug still need some animal tests before human use, the experiments are fewer and safer for the animal because of this initial screen with iPSC.

I told you at the beginning this was going to be astonishing. Induced pluripotent stem cells are a true wonder of science. Not only do they have the power to make authentic, young, energetic heart muscles but they help to uncover the real function of genes and find new drugs for heart disease. And the same is true for every organ in the body.

This is how the future looks. Remote monitoring creates for you a "digital twin" with the data stream from your life: your vital signs, genetic makeup, social circumstances, and exposures to external risks.[23] As with aircraft components now, a real-time computer model of your health anticipates and deflects approaching threats. Your heart is recreated on a tiny microfluidic chip with the multiple cell types all derived from your own cells, reprogrammed into iPSCs.[24] When your digital twin sends you for treatment, the ideal drug for you can be selected by testing on your personalized chip. Or it may indicate that one of your genes needs to be tweaked by some CRISPR/Cas9 gene therapy. You may be given a temporary heart support device to allow your heart to recover. Maybe by then we will have built a replacement heart with sophisticated new robotics plus 3D bioprinted iPSC cells. But we don't underestimate the new challenges the heart can present in this journey. Since your own heart has had 520 million years of evolution to get things right, we still have a bit of catching up to do.[25]

ACKNOWLEDGMENTS

I would like to thank Jaime Marshall, my agent, for his perspicacious input and patient help in coaching me on my first steps into writing for a nonacademic audience. I'm grateful to my editor Matt Browne for intelligent and supportive shaping of the manuscript, Catherine Wilson for imaginative additions to the text, and Matt Holford for the inventive illustrations. There are many, many colleagues, postdocs, and students that I would like to thank, especially Peter O'Gara (the myocyte maestro), Julia Gorelik (the nanodomain queen), and Alex Lyon (the Takotsubo guru). I'm also eternally grateful to my husband Ray for his constant support and expert technical advice.

CHAPTER 1

1. J. Miller, *The Body in Question* (London: Random House, 1980).

2. H. Kato, A. B. Jena, and Y. Tsugawa, "Patient Mortality after Surgery on the Surgeon's Birthday: Observational Study," *BMJ* 371 (2020): m4381; B. N. Greenwood, S. Carnahan, and L. Huang, "Patient-Physician Gender Concordance and Increased Mortality among Female Heart Attack Patients," *Proceedings of the National Academy of Sciences USA* 115 (2018): 8569–8574.

3. R. Sinharay, J. Gong, B. Barratt, P. Ohman-Strickland, S. Ernst, F. J. Kelly, J. J. Zhang, P. Collins, P. Cullinan, and K. F. Chung, "Respiratory and Cardiovascular Responses to Walking Down a Traffic-Polluted Road Compared with Walking in a Traffic-Free Area in Participants Aged 60 years and Older with Chronic Lung or Heart Disease and Age-Matched Healthy Controls: A Randomised, Crossover Study," *Lancet* 391 (2018): 339–349.

CHAPTER 2

1. "Global Health Estimates," World Health Organization, accessed December 21, 2021, https://www.who.int/data/global-health-es timates.

2. E. J. Benjamin, P. Muntner, A. Alonso, M. S. Bittencourt, C. W. Callaway, A. P. Carson, A. M. Chamberlain, A. R. Chang, S. Cheng, and S. R. Das et al., *Heart Disease and Stroke Statistics—2019 Update: A Report From the American Heart Association*, circulation 139 (2019): e56–e528, doi: 10.1161/CIR.0000000000000659.

3. "Heart Statistics," British Heart Foundation, accessed December 21, 2021, https://www.bhf.org.uk/what-we-do/our-research/heart -statistics.

4. M. Sims, R. Maxwell, L. Bauld, and A. Gilmore, "Short Term Impact of Smoke-Free Legislation in England: Retrospective Analysis of Hospital Admissions for Myocardial Infarction," *BMJ* 340 (2010): c2161.

5. R. Chou, T. Dana, I. Blazina, M. Daeges, and T. L. Jeanne, "Statins for Prevention of Cardiovascular Disease in Adults: Evidence Report and Systematic Review for the US Preventive Services Task Force," *Journal of the American Medical Association* 316 (2016): 2008–2024.

6. C. J. Taylor, J. M. Ordóñez-Mena, N. R. Jones, A. K. Roalfe, S. Lay-Flurrie, T. Marshall, and F. D. R. Hobbs, "National Trends in Heart Failure Mortality in Men and Women, United Kingdom, 2000–2017," *European Journal of Heart Failure* 23 (2021): 3–12; G. A. Mensah, G. S. Wei, P. D. Sorlie, L. J. Fine, Y. Rosenberg, P. G. Kaufmann, M. E. Mussolino, L. L. Hsu, E. Addou, and M. M. Engelgau et al., "Decline in Cardiovascular Mortality: Possible Causes and Implications," *Circulation Research* 120 (2017): 366–380.

7. E. C. Matthay, K. A. Duchowny, A. R. Riley, and S. Galea, "Projected All-Cause Deaths Attributable to COVID-19-Related Unemployment in the United States," *American Journal of Public Health* 111 (2021): 696–699; A. N. S. Cartaxo, F. I. C. Barbosa, P. H. de Souza Bermejo, M. F. Moreira, and D. N. Prata, "The Exposure Risk to COVID-19 in Most Affected Countries: A Vulnerability Assessment Model," *PLOS One* 16 (2021): e0248075.

8. R. Chetty, M. Stepner, S. Abraham, S. Lin, B. Scuderi, N. Turner, A. Bergeron, and D. Cutler, "The Association between Income and Life Expectancy in the United States, 2001–2014," *Journal of the American Medical Association* 315 (2016): 1750–1766.

9. "Homeless People Die 30 Years Younger, Study Suggests," BBC News, December 21, 2011, https://www.bbc.co.uk/news/uk-162 72120.

10. M. R. Sterling, J. B. Ringel, L. C. Pinheiro, M. M. Safford, E. B. Levitan, E. Phillips, T. M. Brown, and P. Goyal, "Social Determinants of Health and 90-Day Mortality after Hospitalization for Heart Failure in the REGARDS Study," *Journal of the American Heart Association* 9 (2020): e014836.

11. M. G. Marmot, G. D. Smith, S. Stansfeld, C. Patel, F. North, J. Head, I. White, E. Brunner, and A. Feeney, "Health Inequalities among British Civil Servants: The Whitehall II Study," *Lancet* 337 (1991) 1387–1393.

12. A. Tanaka, M. J. Shipley, C. A. Welch, N. E. Groce, M. G. Marmot, M. Kivimaki, A. Singh-Manoux, and E. J. Brunner, "Socioeconomic Inequality in Recovery from Poor Physical and Mental Health in Mid-life and Early Old Age: Prospective Whitehall II Cohort Study," *Journal of Epidemiology and Community Health* 72 (2018): 309–313.

13. M. Razzoli, K. Nyuyki-Dufe, A. Gurney, C. Erickson, J. McCallum, N. Spielman, M. Marzullo, J. Patricelli, M. Kurata, and E. A. Pope et al., "Social Stress Shortens Lifespan in Mice," *Aging Cell* 17 (2018): e12778.

14. S. Schafer, A. de Marvao, E. Adami, L. R. Fiedler, B. Ng, E. Khin, O. J. Rackham, S. van Heesch, C. J. Pua, and M. Kui et al., "Titin-Truncating Variants Affect Heart Function in Disease Cohorts and the General Population," *Nature Genetics* 49 (2017): 46–53.

15. P. Garcia-Pavia, Y. Kim, M. A. Restrepo-Cordoba, I. G. Lunde, H. Wakimoto, A. M. Smith, C. N. Toepfer, K. Getz, J. Gorham, and P. Patel et al., "Genetic Variants Associated with Cancer Therapy-Induced Cardiomyopathy," *Circulation* 140 (2019): 31–41.

16. J. S. Ware, A. Amor-Salamanca, U. Tayal, R. Govind, I. Serrano, J. Salazar-Mendiguchía, J. M. García-Pinilla, D. A. Pascual-Figal, J. Nuñez, and G. Guzzo-Merello et al., "Genetic Etiology for Alcohol-Induced Cardiac Toxicity," *Journal of the American College of Cardiology* 71 (2018): 2293–2302.

17. Ware et al.,"Genetic Etiology."

18. J. S. Ware, J. Li, E. Mazaika, C. M. Yasso, T. DeSouza, T. P. Cappola, E. J. Tsai, D. Hilfiker-Kleiner, C. A. Kamiya, and F. Mazzarotto et al., "Shared Genetic Predisposition in Peripartum and Dilated Cardiomyopathies," *New England Journal of Medicine* 374 (2016): 233–241.

19. S. P. Murphy, N. E. Ibrahim, and J. L. Januzzi Jr., "Heart Failure with Reduced Ejection Fraction: A Review," *Journal of the American Medical Association* 324 (2020): 488–504.

20. S. Sattler, P. Fairchild, F. M. Watt, N. Rosenthal, and S. E. Harding, "The Adaptive Immune Response to Cardiac Injury: The True Roadblock to Effective Regenerative Therapies?," *NPJ Regenerative Medicine* 2 (2017): 19.

21. Benjamin et al., *Heart Disease and Stroke Statistics*.

22. "Breast Cancer: Survival," Cancer Research UK, last reviewed January 3, 2020, https://www.cancerresearchuk.org/about-cancer /breast-cancer/survival.

23. D. Slamon, W. Eiermann, N. Robert, T. Pienkowski, M. Martin, M. Press, J. Mackey, J. Glaspy, A. Chan, and M. Pawlicki et al., "Adjuvant Trastuzumab in HER2-Positive Breast Cancer," *New England Journal of Medicine* 365 (2011): 1273–1283.

24. Slamon et al., "Adjuvant Trastuzumab."

25. M. J. Feinstein, M. Bogorodskaya, G. S. Bloomfield, R. Vedanthan, M. J. Siedner, G. F. Kwan, and C. T. Longenecker, "Cardiovascular Complications of HIV in Endemic Countries," *Current Cardiology Reports* 18 (2016): 113.

CHAPTER 3

1. N. Nair, "Epidemiology and Pathogenesis of Heart Failure with Preserved Ejection Fraction," *Reviews in Cardiovascular Medicine* 21 (2020): 531–540.

2. M. J. Lab, A. Bhargava, P. T. Wright, and J. Gorelik, "The Scanning Ion Conductance Microscope for Cellular Physiology," *American Journal of Physiology-Heart and Circulatory Physiology* 304 (2013): H1–11.

3. B. N. Pham and S. V. Chaparro, "Left Ventricular Assist Device Recovery: Does Duration of Mechanical Support Matter?," *Heart Failure Reviews* 24 (2019): 237–244.

CHAPTER 4

1. R. Plomin, *Blueprint: How DNA Makes Who We Are* (Cambridge, MA: MIT Press, 2018).

2. A. V. Khera, M. Chaffin, K. G. Aragam, M. E. Haas, C. Roselli, S. H. Choi, P. Natarajan, E. S. Lander, S. A. Lubitz, and P. T. Ellinor et al., "Genome-wide Polygenic Scores for Common Diseases Identify Individuals with Risk Equivalent to Monogenic Mutations," *Nature Genetics* 50 (2018): 1219–1224.

3. A. Kaura, V. Panoulas, B. Glampson, J. Davies, A. Mulla, K. Woods, J. Omigie, A. D. Shah, K. M. Channon, and J. N. Weber et al., "Association of Troponin Level and Age with Mortality in 250,000 Patients: Cohort Study across Five UK Acute Care Centres," *BMJ* 367 (2019): l6055.

4. S. S. Mahmood, D. Levy, R. S. Vasan, and T. J. Wang, "The Framingham Heart Study and the Epidemiology of Cardiovascular Disease: A Historical Perspective," *Lancet* 383 (2014): 999–1008.

5. UK Biobank, accessed December 21, 2021, https://www.ukbiobank.ac.uk/.

6. P. J. Landrigan, R. Fuller, N. J. R. Acosta, O. Adeyi, R. Arnold, N. N. Basu, A. B. Baldé, R. Bertollini, S. Bose-O'Reilly, and J. I. Boufford et al., "The Lancet Commission on Pollution and Health," *Lancet* 391 (2018): 462–512.

7. H. Byrnes, "The 30 Most Polluted Places in America," *24/7 Wall St*, last updated January 6, 2020, https://247wallst.com/special-report/2019/12/03/the-30-most-polluted-places-in-america/.

8. Sinharay, "Respiratory and Cardiovascular Responses," 339–349.

9. W. von Rosenberg, T. Chanwimalueang, V. Goverdovsky, N. S. Peters, C. Papavassiliou, and D. P. Mandic, "Hearables: Feasibility of Recording Cardiac Rhythms from Head and In-ear Locations," *Royal Society Open Science* 4 (2017): 171214.

10. M. Van Puyvelde, X. Neyt, F. McGlone, and N. Pattyn, "Voice Stress Analysis: A New Framework for Voice and Effort in Human Performance," *Frontiers in Psychology* 9 (2018): 1994.

11. Y. Yan, X. Ma, L. Yao, and J. Ouyang, "Noncontact Measurement of Heart Rate Using Facial Video Illuminated under Natural Light and Signal Weighted Analysis," *Bio-Medical Materials and Engineering* 26, Suppl. 1 (2015): S903–909.

12. M. Litviňuková, C. Talavera-López, H. Maatz, D. Reichart, C. L. Worth, E. L. Lindberg, M. Kanda, K. Polanski, M. Heinig, and M. Lee et al., "Cells of the Adult Human Heart," *Nature* 588 (2020): 466–472.

13. D. P. O'Regan, "Putting Machine Learning into Motion: Applications in Cardiovascular Imaging," *Clinical Radiology* 75 (2020): 33–37.

14. S. Schafer, A. de Marvao, E. Adami, L. R. Fiedler, B. Ng, E. Khin, O. J. Rackham, S. van Heesch, C. J. Pua, and M. Kui et al., "Titin-Truncating Variants Affect Heart Function in Disease Cohorts and the General Population," *Nature Genetics* 49 (2017): 46–53.

15. G. A. Bello, T. J. W. Dawes, J. Duan, C. Biffi, A. de Marvao, L. Howard, J. S. R. Gibbs, M. R. Wilkins, S. A. Cook, and D. Rueckert et al., "Deep Learning Cardiac Motion Analysis for Human Survival Prediction," *Nature Machine Intelligence* 1 (2019): 95–104.

16. "Amazon Scrapped 'Sexist AI' Tool," BBC News, October 10, 2018.

CHAPTER 5

1. A. R. Orkaby, J. A. Driver, Y. L. Ho, B. Lu, L. Costa, J. Honerlaw, A. Galloway, J. L. Vassy, D. E. Forman, and J. M. Gaziano et al., "Association of Statin Use with All-Cause and Cardiovascular Mortality in US Veterans 75 Years and Older," *Journal of the American Medical Association* 324 (2020): 68–78.

2. O. Bergmann, S. Zdunek, A. Felker, M. Salehpour, K. Alkass, S. Bernard, S. L. Sjostrom, M. Szewczykowska, T. Jackowska, and C.

Dos Remedios et al., "Dynamics of Cell Generation and Turnover in the Human Heart," *Cell* 161 (2015): 1566–1575.

3. K. L. Spalding, O. Bergmann, K. Alkass, S. Bernard, M. Salehpour, H. B. Huttner, E. Boström, I. Westerlund, C. Vial, and B. A. Buchholz et al., "Dynamics of Hippocampal Neurogenesis in Adult Humans," *Cell* 153 (2013): 1219–1227.

4. M. Didié, D. Biermann, R. Buchert, A. Hess, K. Wittköpper, P. Christalla, S. Döker, F. Jebran, F. Schöndube, and H. Reichenspurner et al., "Preservation of Left Ventricular Function and Morphology in Volume-Loaded versus Volume-Unloaded Heterotopic Heart Transplants," *American Journal of Physiology-Heart and Circulatory Physiology* 305 (2013): H533–541.

5. B. N. Pham and S. V. Chaparro, "Left Ventricular Assist Device Recovery: Does Duration of Mechanical Support Matter?," *Heart Failure Reviews* 24 (2019): 237–244.

6. D. W. T. Wundersitz, B. A. Gordon, C. J. Lavie, V. Nadurata, and M. I. C. Kingsley, "Impact of Endurance Exercise on the Heart of Cyclists: A Systematic Review and Meta-analysis," *Progress in Cardiovascular Diseases* 63 (2020): 750–761.

7. "Heart Statistics," British Heart Foundation, accessed December 21, 2021, https://www.bhf.org.uk/what-we-do/our-research/heart-statistics; "High Blood Pressure," Centers for Disease Control and Prevention (CDC), last reviewed October 22, 2020, https://www.cdc.gov/bloodpressure.

8. R. M. Carey and P. K. Whelton, "The 2017 American College of Cardiology/American Heart Association Hypertension Guideline: A Resource for Practicing Clinicians," *Annals of Internal Medicine* 168 (2018): 359–360.

9. C. Maraboto and K. C. Ferdinand, "Update on Hypertension in African-Americans," *Progress in Cardiovascular Diseases* 63 (2020): 33–39.

10. M. P. Gupta, "Factors Controlling Cardiac Myosin-Isoform Shift during Hypertrophy and Heart Failure," *Journal of Molecular and Cellular Cardiology* 43 (2007): 388–403.

11. M. Shanmuganathan, J. Vughs, M. Noseda, and C. Emanueli, "Exosomes: Basic Biology and Technological Advancements Suggesting Their Potential as Ischemic Heart Disease Therapeutics," *Frontiers in Physiology* 9 (2018): 1159.

12. Shanmuganathan et al., "Exosomes."

13. R. Tikhomirov, B. R. Donnell, F. Catapano, G. Faggian, J. Gorelik, F. Martelli and C. Emanueli, "Exosomes: From Potential Culprits to New Therapeutic Promise in the Setting of Cardiac Fibrosis," *Cells* 9 (2020): 592.

14. Tikhomirov et al., "Exosomes."

CHAPTER 6

1. J. L. Ardell and J. A. Armour, "Neurocardiology: Structure-Based Function," *Comprehensive Physiology* 6 (2016): 1635–1653.

2. S. N. Garfinkel and H. D. Critchley, "Threat and the Body: How the Heart Supports Fear Processing," *Trends in Cognitive Science* 20 (2016): 34–46.

3. S. W. Jeong, S. H. Kim, S. H. Kang, H. J. Kim, C. H. Yoon, T. J. Youn, and I. H. Chae, "Mortality Reduction with Physical Activity in Patients with and without Cardiovascular Disease," *European Heart Journal* 40 (2019): 3547–3555.

CHAPTER 7

1. Y. J. Akashi, H. M. Nef, and A. R. Lyon, "Epidemiology and Pathophysiology of Takotsubo Syndrome," *Nature Reviews Cardiology* 12 (2015): 387–397.

2. Akashi, Nef, and Lyon, "Epidemiology and Pathophysiology."

3. Y. H. Shams, K. Feldt, and M. Stålberg, "A Missed Penalty Kick Triggered Coronary Death in the Husband and Broken Heart Syndrome in the Wife," *American Journal of Cardiology* 116 (2015): 1639–1642.

4. W. Kirkup and D. W. Merrick, "A Matter of Life and Death: Population Mortality and Football Results," *Journal of Epidemiology and Community Health* 57 (2003): 429–432.

5. D. Carroll, S. Ebrahim, K. Tilling, J. Macleod, and G. D. Smith, "Admissions for Myocardial Infarction and World Cup Football: Database Survey," *BMJ* 325 (2002): 1439–1442.

6. U. Wilbert-Lampen, D. Leistner, S. Greven, T. Pohl, S. Sper, C. Völker, D. Güthlin, A. Plasse, A. Knez, and H. Küchenhoff et al., "Cardiovascular Events during World Cup Soccer," *New England Journal of Medicine* 358 (2008): 475–483.

7. M. A. Kaballo, A. Yousif, A. M. Abdelrazig, A. A. Ibrahim, and T. G. Hennessy, "Takotsubo Cardiomyopathy after a Dancing Session: A Case Report," *Journal of Medical Case Reports* 5 (2011): 533.

8. J. R. Ghadri, A. Sarcon, J. Diekmann, D. R. Bataiosu, V. L. Cammann, S. Jurisic, L. C. Napp, M. Jaguszewski, F. Scherff, and P. Brugger et al., "Happy Heart Syndrome: Role of Positive Emotional Stress in Takotsubo Syndrome," *European Heart Journal* 37 (2016): 2823–2829.

9. C. E. Rodriguez-Castro, F. Saifuddin, M. Porres-Aguilar, S. Said, D. Gough, T. Siddiqui, D. Mukherjee, and A. Abbas, "Reverse Takotsubo Cardiomyopathy with Use of Male Enhancers," *Baylor University Medical Center Proceedings* 28 (2015): 78–80.

10. A. Ali, A. K. Niazi, P. Minko, P. J. Saha, K. Elliott, N. Bhatnagar and S. Ayad, "A Case of Takotsubo Cardiomyopathy after Local Anesthetic and Epinephrine Infiltration," *Cureus* 10 (2018): e3173.

11. A. Singh, T. Sturzoiu, S. Vallabhaneni, and J. Shirani, "Stress Cardiomyopathy Induced during Dobutamine Stress Echocardiography," *International Journal of Critical Illness and Injury Science* 10 (2020): 43–48.

12. L. E. Gibson, M. R. Klinker, and M. J. Wood, "Variants of Takotsubo Syndrome in the Perioperative Period: A Review of Potential Mechanisms and Anaesthetic Implications," *Anaesthesia Critical Care and Pain Medicine* 39 (2020): 647–654.

13. F. Y. Marmoush, M. F. Barbour, T. E. Noonan, and M. O. Al-Qadi, "Takotsubo Cardiomyopathy: A New Perspective in Asthma," *Case Reports in Cardiology* (2015): 640795.

14. V. Zvonarev, "Takotsubo Cardiomyopathy: Medical and Psychiatric Aspects. Role of Psychotropic Medications in the Treatment of Adults with 'Broken Heart' Syndrome," *Cureus* 11 (2019): e5177.

15. O. A. Kajander, M. P. Virtanen, S. Sclarovsky, and K. C. Nikus, "Iatrogenic Inverted Takotsubo Syndrome Following Intravenous Adrenaline Injections for an Allergic Reaction," *International Journal of Cardiology* 165 (2013): e3–5.

16. A. Aweimer, I. El-Battrawy, I. Akin, M. Borggrefe, A. Mügge, P. C. Patsalis, A. Urban, M. Kummer, S. Vasileva, and A. Stachon et al., "Abnormal Thyroid Function Is Common in Takotsubo Syndrome and Depends on Two Distinct Mechanisms: Results of a Multicentre Observational Study," *Journal of Internal Medicine* 289 (2021): 675–687; S. Cappelletti, C. Ciallella, M. Aromatario, H. Ashrafian, S. Harding, and T. Athanasiou, "Takotsubo Cardiomyopathy and Sepsis," *Angiology* 68 (2017): 288–303.

17. N. A. Morris, A. Chatterjee, O. L. Adejumo, M. Chen, A. E. Merkler, S. B. Murthy, and H. Kamel, "The Risk of Takotsubo Cardiomyopathy in Acute Neurological Disease," *Neurocritical Care* 30 (2019): 171–176.

18. Marmoush et al., "Takotsubo Cardiomyopathy."

19. H. Gong, D. L. Adamson, H. K. Ranu, W. J. Koch, J. F. Heubach, U. Ravens, O. Zolk, and S. E. Harding, "The Effect of Gi-Protein Inactivation on Basal, and Beta(1)- and Beta(2)AR-Stimulated Contraction of Myocytes from Transgenic Mice Overexpressing the Beta(2)-Adrenoceptor," *British Journal of Pharmacology* 131 (2000): 594–600.

20. A. R. Hasseldine, E. A. Harper, and J. W. Black, "Cardiac-specific Overexpression of Human Beta2 Adrenoceptors in Mice Exposes Coupling to Both Gs and Gi Proteins," *British Journal of Pharmacology* 138 (2003): 1358–1366.

21. H. Paur, P. T. Wright, M. B. Sikkel, M. H. Tranter, C. Mansfield, P. O'Gara, D. J. Stuckey, V. O. Nikolaev, I. Diakonov, and L. Pannell et al., "High Levels of Circulating Epinephrine Trigger Apical Cardiodepression in a β2-adrenergic Receptor/Gi-dependent

Manner: A New Model of Takotsubo Cardiomyopathy," *Circulation* 126 (2012): 697–706.

22. Paur et al., "High Levels."

23. Y. J. Akashi, H. M. Nef, and A. R. Lyon, "Epidemiology and Pathophysiology of Takotsubo Syndrome," *Nature Reviews Cardiology* 12 (2015): 387–397.

24. S. E. Harding and H. Gong, "Beta-Adrenoceptor Blockers as Agonists: Coupling of Beta2-Adrenoceptors to Multiple G-Proteins in the Failing Human Heart," *Congestive Heart Failure* 10 (2004): 181–185.

25. F. Waagstein, K. Caidahl, I. Wallentin, C. H. Bergh, and A. Hjalmarson, "Long-Term Beta-Blockade in Dilated Cardiomyopathy: Effects of Short- and Long-Term Metoprolol Treatment Followed by Withdrawal and Readministration of Metoprolol," *Circulation* 80 (1989): 551–563.

26. L. S. Couch, J. Fiedler, G. Chick, R. Clayton, E. Dries, L. M. Wienecke, L. Fu, J. Fourre, P. Pandey, and A. A. Derda et al., "Circulating MicroRNAs Predispose to Takotsubo Syndrome Following High-Dose Adrenaline Exposure," *Cardiovascular Research* (2021), doi: 10.1093/cvr/cvab210.

CHAPTER 8

1. L. Mosca, E. Barrett-Connor, and N. K. Wenger, "Sex/Gender Differences in Cardiovascular Disease Prevention: What a Difference a Decade Makes," *Circulation* 124 (2011): 2145–2154.

2. F. Mauvais-Jarvis, N. Bairey Merz, P. J. Barnes, R. D. Brinton, J. J. Carrero, D. L. DeMeo, G. J. De Vries, C. N. Epperson, R. Govindan, and S. L. Klein et al., "Sex and Gender: Modifiers of Health, Disease, and Medicine," *Lancet* 396 (2020): 565–582.

3. F. Mauvais-Jarvis et al., "Sex and Gender."

4. K. Matsushita, N. Ding, M. Kou, X. Hu, M. Chen, Y. Gao, Y. Honda, D. Zhao, D. Dowdy, and Y. Mok et al., "The Relationship of COVID-19 Severity with Cardiovascular Disease and Its

Traditional Risk Factors: A Systematic Review and Meta-Analysis," *Global Heart* 15 (2020): 64; I. Torjesen, "Covid-19: Middle Aged Women Face Greater Risk of Debilitating Long Term Symptoms," *BMJ* 372 (2021): n829.

5. Torjesen, "Covid-19."

6. Torjesen.

7. P. J. Connelly, E. Marie Freel, C. Perry, J. Ewan, R. M. Touyz, G. Currie, and C. Delles, "Gender-Affirming Hormone Therapy, Vascular Health and Cardiovascular Disease in Transgender Adults," *Hypertension* 74 (2019): 1266–1274.

8. J. P. Walsh and A. C. Kitchens, "Testosterone Therapy and Cardiovascular Risk," *Trends in Cardiovascular Medicine* 25 (2015): 250–257.

9. M. Gambacciani, A. Cagnacci, and S. Lello, "Hormone Replacement Therapy and Prevention of Chronic Conditions," *Climacteric* 22 (2019): 303–306; A. Cagnacci and M. Venier, "The Controversial History of Hormone Replacement Therapy," *Medicina* (Kaunas, Lithuania) 55 (2019): 602.

10. J. Marsden, "British Menopause Society Consensus Statement: The Risks and Benefits of HRT Before and After a Breast Cancer Diagnosis," *Post Reproductive Health* 25 (2019): 33–37.

11. F. Mauvais-Jarvis, N. Bairey Merz, P. J. Barnes, R. D. Brinton, J. J. Carrero, D. L. DeMeo, G. J. De Vries, C. N. Epperson, R. Govindan, and S. L. Klein et al., "Sex and Gender: Modifiers of Health, Disease, and Medicine," *Lancet* 396 (2020): 565–582.

12. R. O. Roswell, J. Kunkes, A. Y. Chen, K. Chiswell, S. Iqbal, M. T. Roe, and S. Bangalore, "Impact of Sex and Contact-to-Device Time on Clinical Outcomes in Acute ST-Segment Elevation Myocardial Infarction—Findings From the National Cardiovascular Data Registry," *Journal of the American Heart Association* 6 (2017): e004521.

13. M. Erickson, "Why Are So Few Interventional Cardiologists Women? A New Study Offers a Few Clues," *Scope 10K* blog, February 4, 2019, Stanford Medicine, https://scopeblog.stanford.edu

/2019/02/04/why-are-so-few-interventional-cardiologists-women
-a-new-study-offers-a-few-clues/.

14. B. N. Greenwood, S. Carnahan, and L. Huang, "Patient-Physician Gender Concordance and Increased Mortality among Female Heart Attack Patients," *Proceedings of the National Academy of Sciences USA* 115 (2018): 8569–8574.

15. T. A. Laveist and A. Nuru-Jeter, "Is Doctor-Patient Race Concordance Associated with Greater Satisfaction with Care?," *Journal of Health and Social Behavior* 43 (2002): 296–306; T. A. LaVeist, A. Nuru-Jeter, and K. E. Jones, "The Association of Doctor-Patient Race Concordance with Health Services Utilization," *Journal of Public Health Policy* 24 (2003): 312–323.

16. S. Simon and P. M. Ho, "Ethnic and Racial Disparities in Acute Myocardial Infarction," *Current Cardiology Reports* 22 (2020): 88; S. A. Karnati, A. Wee, M. M. Shirke, and A. Harky, "Racial Disparities and Cardiovascular Disease: One Size Fits All Approach?," *Journal of Cardiac Surgery* 35 (2020): 3530–3538.

17. J. J. Chinn, I. K. Martin, and N. Redmond, "Health Equity among Black Women in the United States," *Journal of Women's Health* 30 (2021): 212–219.

18. "Bem Sex Role Inventory" survey, PsyToolkit, accessed December 21, 2021, psytoolkit.org/c/3.4.0/survey?s=BsNnQ.

19. R. Pelletier, N. A. Khan, J. Cox, S. S. Daskalopoulou, M. J. Eisenberg, S. L. Bacon, K. L. Lavoie, K. Daskupta, D. Rabi, and K. H. Humphries et al., "Sex Versus Gender-Related Characteristics: Which Predicts Outcome after Acute Coronary Syndrome in the Young?," *Journal of the American College of Cardiology* 67 (2016): 127–135.

20. A. McGregor, *Sex Matters* (London: Quercus, 2020).

CHAPTER 9

1. N. Bhatia and M. El-Chami, "Leadless Pacemakers: A Contemporary Review," *Journal of Geriatric Cardiology* 15 (2018): 249–253.

2. M. Rav Acha, E. Soifer, and T. Hasin, "Cardiac Implantable Electronic Miniaturized and Micro Devices," *Micromachines* (Basel) 11 (2020): 902.

3. Rav Acha, Soifer, and Hasin, "Cardiac Implantable."

4. Rav Acha, Soifer, and Hasin.

5. Rav Acha, Soifer, and Hasin.

6. L. Bereuter, T. Niederhauser, M. Kucera, D. Loosli, I. Steib, M. Schildknecht, A. Zurbuchen, F. Noti, H. Tanner, and T. Reichlin et al., "Leadless Cardiac Resynchronization Therapy: An In Vivo Proof-of-Concept Study of Wireless Pacemaker Synchronization," *Heart Rhythm* 16 (2019): 936–942.

7. A. B. E. Quast, F. V. Y. Tjong, B. E. Koop, A. A. M. Wilde, R. E. Knops, and M. C. Burke, "Device Orientation of a Leadless Pacemaker and Subcutaneous Implantable Cardioverter-Defibrillator in Canine and Human Subjects and the Effect on Intrabody Communication," *Europace* 20 (2018): 1866–1871.

8. F. M. Merchant, W. C. Levy, and D. B. Kramer, "Time to Shock the System: Moving Beyond the Current Paradigm for Primary Prevention Implantable Cardioverter-Defibrillator Use," *Journal of the American Heart Association* 9 (2020): e015139.

9. Rav Acha, Soifer, and Hasin, "Cardiac Implantable."

10. Rav Acha, Soifer, and Hasin.

11. W. E. Cohn, D. L. Timms, and O. H. Frazier, "Total Artificial Hearts: Past, Present, and Future," *Nature Reviews Cardiology* 12 (2015): 609–617.

12. J. Copeland, S. Langford, J. Giampietro, J. Arancio, and F. Arabia, "Total Artificial Heart Update," *Surgical Technology International* 39 (2021): 243–248.

13. G. Torregrossa, M. Morshuis, R. Varghese, L. Hosseinian, V. Vida, V. Tarzia, A. Loforte, D. Duveau, F. Arabia, and P. Leprince et al., "Results with SynCardia Total Artificial Heart beyond 1 Year," *ASAIO Journal* 60 (2014): 626–634.

14. J. L. Vieira, H. O. Ventura, and M. R. Mehra, "Mechanical Circulatory Support Devices in Advanced Heart Failure: 2020 and Beyond," *Progress in Cardiovascular Diseases* 63 (2020): 630–639.

15. W. E. Cohn, D. L. Timms, and O. H. Frazier, "Total Artificial Hearts: Past, Present, and Future," *Nature Reviews Cardiology* 12 (2015): 609–617.

16. K. K. Khush, L. Potena, W. S. Cherikh, D. C. Chambers, M. O. Harhay, D. Hayes, Jr., E. Hsich, A. Sadavarte, T. P. Singh, and A. Zuckermann et al., "The International Thoracic Organ Transplant Registry of the International Society for Heart and Lung Transplantation: 37th Adult Heart Transplantation Report 2020; Focus on Deceased Donor Characteristics," *Journal of Heart and Lung Transplantation* 39 (2020): 1003–1015.

17. A. F. Sunjaya and A. P. Sunjaya, "Combating Donor Organ Shortage: Organ Care System Prolonging Organ Storage Time and Improving the Outcome of Heart Transplantations," *Cardiovascular Therapeutics* (2019): 9482797, doi: 10.1155/2019/9482797.

18. T. P. K. Fleck, R. Ayala, J. Kroll, M. Siepe, D. Schibilsky, C. Benk, S. Maier, K. Reineker, R. Hoehn, and F. Humburger et al., "Ex-vivo Allograft Perfusion for Complex Pediatric Heart Transplant Recipients," *Annals of Thoracic Surgery* 112 (2021): 1275–1280.

19. F. D. Pagani, "Use of Heart Donors Following Circulatory Death: A Viable Addition to the Heart Donor Pool," *Journal of the American College of Cardiology* 73 (2019): 1460–1462.

CHAPTER 10

1. K. O'Donoghue, J. Chan, J. de la Fuente, N. Kennea, A. Sandison, J. R. Anderson, I. A. Roberts, and N. M. Fisk, "Microchimerism in Female Bone Marrow and Bone Decades after Fetal Mesenchymal Stem-Cell Trafficking in Pregnancy," *Lancet* 364 (2004): 179–182.

2. S. Nguyen Huu, M. Oster, S. Uzan, F. Chareyre, S. Aractingi, and K. Khosrotehrani, "Maternal Neoangiogenesis during Pregnancy Partly Derives from Fetal Endothelial Progenitor Cells," *Proceedings of the National Academy of Sciences USA* 104 (2007): 1871–1876.

3. K. O'Donoghue, H. A. Sultan, F. A. Al-Allaf, J. R. Anderson, J. Wyatt-Ashmead, and N. M. Fisk, "Microchimeric Fetal Cells Cluster at Sites of Tissue Injury in Lung Decades after Pregnancy," *Reproductive Biomedicine Online* 16 (2008): 382–390.

4. C. E. Eckfeldt, E. M. Mendenhall, and C. M. Verfaillie, "The Molecular Repertoire of the 'Almighty' Stem Cell," *Nature Reviews Molecular Cell Biology* 6 (2005): 726–737.

5. See, for example, Future Health Biobank, accessed December 21, 2021, https://futurehealthbiobank.com/.

6. B. J. Haubner, J. Schneider, U. Schweigmann, T. Schuetz, W. Dichtl, C. Velik-Salchner, J. I. Stein, and J. M. Penninger, "Functional Recovery of a Human Neonatal Heart after Severe Myocardial Infarction," *Circulation Research* 118 (2016): 216–221.

7. P. Menasché, O. Alfieri, S. Janssens, W. McKenna, H. Reichenspurner, L. Trinquart, J. T. Vilquin, J. P. Marolleau, B. Seymour, and J. Larghero et al., "The Myoblast Autologous Grafting in Ischemic Cardiomyopathy (MAGIC) Trial: First Randomized Placebo-Controlled Study of Myoblast Transplantation," *Circulation* 117 (2008): 1189–1200.

8. M. Gyöngyösi, P. M. Haller, D. J. Blake, and E. Martin Rendon, "Meta-Analysis of Cell Therapy Studies in Heart Failure and Acute Myocardial Infarction," *Circulation Research* 123 (2018): 301–308.

9. Menasché et al., "Myoblast Autologous Grafting."

10. M. Vasa, S. Fichtlscherer, A. Aicher, K. Adler, C. Urbich, H. Martin, A. M. Zeiher, and S. Dimmeler, "Number and Migratory Activity of Circulating Endothelial Progenitor Cells Inversely Correlate with Risk Factors for Coronary Artery Disease," *Circulation Research* 89 (2001): E1–7.

11. J. Vasanthan, N. Gurusamy, S. Rajasingh, V. Sigamani, S. Kirankumar, E. L. Thomas, and J. Rajasingh, "Role of Human Mesenchymal Stem Cells in Regenerative Therapy," *Cells* 10 (2020): 54.

12. K. Wei, V. Serpooshan, C. Hurtado, M. Diez-Cuñado, M. Zhao, S. Maruyama, W. Zhu, G. Fajardo, M. Noseda, and K. Nakamura

et al., "Epicardial FSTL1 Reconstitution Regenerates the Adult Mammalian Heart," *Nature* 525 (2015): 479–485.

13. K. Wei et al., "Epicardial FSTL1 Reconstitution."

14. Y. Lin and J. Zou, "Differentiation of Cardiomyocytes from Human Pluripotent Stem Cells in Fully Chemically Defined Conditions," *STAR Protocol* 1 (2020), doi: 10.1016/j.xpro.2020.100015.

15. See, for example, iCell Cardiomyocytes products page, Fujifilm Cellular Dynamics, accessed December 21, 2021, https://www.fujifilmcdi.com/products/cardiac-cells/icell-cardiomyocytes.

16. M. N. Hirt, J. Boeddinghaus, A. Mitchell, S. Schaaf, C. Börnchen, C. Müller, H. Schulz, N. Hubner, J. Stenzig, and A. Stoehr et al., "Functional Improvement and Maturation of Rat and Human Engineered Heart Tissue by Chronic Electrical Stimulation," *Journal of Molecular and Cellular Cardiology* 74 (2014): 151–161.

CHAPTER 11

1. A. Mathur, F. Fernández-Avilés, J. Bartunek, A. Belmans, F. Crea, S. Dowlut, M. Galiñanes, M. C. Good, J. Hartikainen, and C. Hauskeller et al., "The Effect of Intracoronary Infusion of Bone Marrow-Derived Mononuclear Cells on All-Cause Mortality in Acute Myocardial Infarction: The BAMI Trial," *European Heart Journal* 41 (2020): 3702–3710.

2. B. P. Halliday, R. Wassall, A. S. Lota, Z. Khalique, J. Gregson, S. Newsome, R. Jackson, T. Rahneva, R. Wage, and G. Smith et al., "Withdrawal of Pharmacological Treatment for Heart Failure in Patients with Recovered Dilated Cardiomyopathy (TRED-HF): An Open-Label, Pilot, Randomised Trial," *Lancet* 393 (2019): 61–73.

3. E. J. Birks, S. G. Drakos, S. R. Patel, B. D. Lowes, C. H. Selzman, R. C. Starling, J. Trivedi, M. S. Slaughter, P. Alturi, and D. Goldstein et al., "Prospective Multicenter Study of Myocardial Recovery Using Left Ventricular Assist Devices (RESTAGE-HF [Remission from Stage D Heart Failure]): Medium-Term and Primary End Point Results," *Circulation* 142 (2020): 2016–2028.

4. L. Liberale, F. Montecucco, L. Schwarz, T. F. Lüscher, and G. G. Camici, "Inflammation and Cardiovascular Diseases: Lessons from Seminal Clinical Trials," *Cardiovascular Research* 117 (2021): 411–422.

5. F. Ferrua and A. Aiuti, "Twenty-Five Years of Gene Therapy for ADA-SCID: From Bubble Babies to an Approved Drug," *Human Gene Therapy* 28 (2017): 972–981.

6. A. Cantore, M. Ranzani, C. C. Bartholomae, M. Volpin, P. D. Valle, F. Sanvito, L. S. Sergi, P. Gallina, F. Benedicenti, and D. Bellinger et al., "Liver-Directed Lentiviral Gene Therapy in a Dog Model of Hemophilia B," *Science Translational Medicine* 7 (2015): 277ra228.

7. A. Cantore and L. Naldini, "WFH State-of-the-Art Paper 2020: In Vivo Lentiviral Vector Gene Therapy for Haemophilia," *Haemophilia* S3 (2021): 122–125.

8. F. del Monte, R. J. Hajjar, and S. E. Harding, "Overwhelming Evidence of the Beneficial Effects of SERCA Gene Transfer in Heart Failure," *Circulation Research* 88 (2001): E66–67.

9. M. B. Sikkel, C. Hayward, K. T. MacLeod, S. E. Harding, and A. R. Lyon, "SERCA2a Gene Therapy in Heart Failure: An Anti-Arrhythmic Positive Inotrope," *British Journal of Pharmacology* (2014): 171, 38–54.

10. B. E. Jaski, M. L. Jessup, D. M. Mancini, T. P. Cappola, D. F. Pauly, B. Greenberg, K. Borow, H. Dittrich, K. M. Zsebo, and R. J. Hajjar, "Calcium Upregulation by Percutaneous Administration of Gene Therapy in Cardiac Disease (CUPID Trial): A First-in-Human Phase 1/2 Clinical Trial," *Journal of Cardiac Failure* 15 (2009): 171–181.

11. B. Greenberg, J. Butler, G. M. Felker, P. Ponikowski, A. A. Voors, A. S. Desai, D. Barnard, A. Bouchard, B. Jaski, and A. R. Lyon et al., "Calcium Upregulation by Percutaneous Administration of Gene Therapy in Patients with Cardiac Disease (CUPID 2): A Randomised, Multinational, Double-Blind, Placebo-Controlled, Phase 2b Trial," *Lancet* 387 (2016): 1178–1186.

12. A. R. Lyon, D. Babalis, A. C. Morley-Smith, M. Hedger, A. Suarez Barrientos, G. Foldes, L. S. Couch, R. A. Chowdhury, K. N. Tzortzis, and N. S. Peters et al., "Investigation of the Safety and Feasibility of AAV1/SERCA2a Gene Transfer in Patients with Chronic Heart Failure Supported with a Left Ventricular Assist Device—the SERCA-LVAD TRIAL," *Gene Therapy* 27 (2020): 579–590.

13. Lyon et al., "AAV1/SERCA2a Gene Transfer."

14. K. Gabisonia, G. Prosdocimo, G. D. Aquaro, L. Carlucci, L. Zentilin, I. Secco, H. Ali, L. Braga, N. Gorgodze, and F. Bernini et al., "MicroRNA Therapy Stimulates Uncontrolled Cardiac Repair after Myocardial Infarction in Pigs," *Nature* 569 (2019): 418–422.

15. L. M. Gan, M. Lagerström-Fermér, L. G. Carlsson, C. Arfvidsson, A. C. Egnell, A. Rudvik, M. Kjaer, A. Collén, J. D. Thompson, and J. Joyal et al., "Intradermal Delivery of Modified mRNA Encoding VEGF-A in Patients with Type 2 Diabetes," *Nature Communications* 10 (2019): 871.

16. Y. Shiba, T. Gomibuchi, T. Seto, Y. Wada, H. Ichimura, Y. Tanaka, T. Ogasawara, K. Okada, N. Shiba, and K. Sakamoto et al., "Allogeneic Transplantation of iPS Cell-Derived Cardiomyocytes Regenerates Primate Hearts," *Nature* 538 (2016): 388–391.

17. L. Gao, Z. R. Gregorich, W. Zhu, S. Mattapally, Y. Oduk, X. Lou, R. Kannappan, A. V. Borovjagin, G. P. Walcott, and A. E. Pollard et al., "Large Cardiac Muscle Patches Engineered from Human Induced-Pluripotent Stem Cell-Derived Cardiac Cells Improve Recovery from Myocardial Infarction in Swine," *Circulation* 137 (2018): 1712–1730.

18. P. Menasché, V. Vanneaux, A. Hagège, A. Bel, B. Cholley, A. Parouchev, I. Cacciapuoti, R. Al-Daccak, N. Benhamouda, and H. Blons et al., "Transplantation of Human Embryonic Stem Cell-Derived Cardiovascular Progenitors for Severe Ischemic Left Ventricular Dysfunction," *Journal of the American College of Cardiology* 71 (2018): 429–438.

19. Smriti Mallapaty, "Revealed: Two Men in China Were First to Receive Pioneering Stem-Cell Treatment for Heart Disease," *Nature*,

May 13, 2020, https://www.nature.com/articles/d41586-020-01285 -w. See also "Osaka University Transplants iPS Cell-based Heart Cells in World's First Clinical Trial," *Japan Times*, January 28, 2020, https://www.japantimes.co.jp/news/2020/01/28/national/science -health/osaka-university-transplants-ips-cell-based-heart-cells-worlds -first-clinical-trial/.

20. J. G. W. Smith, T. Owen, J. R. Bhagwan, D. Mosqueira, E. Scott, I. Mannhardt, A. Patel, R. Barriales-Villa, L. Monserrat, and A. Hansen et al., "Isogenic Pairs of hiPSC-CMs with Hypertrophic Cardiomyopathy/LVNC-Associated ACTC1 E99K Mutation Unveil Differential Functional Deficits," *Stem Cell Reports* 11 (2018): 1226–1243.

21. G. Duncan, K. Firth, V. George, M. D. Hoang, A. Staniforth, G. Smith, and C. Denning, "Drug-Mediated Shortening of Action Potentials in LQTS2 Human Induced Pluripotent Stem Cell-Derived Cardiomyocytes," *Stem Cells and Development* 26 (2017): 1695–1705.

22. N. Sayed, C. Liu, M. Ameen, F. Himmati, J. Z. Zhang, S. Khanamiri, J. R. Moonen, A. Wnorowski, L. Cheng, and J. W. Rhee et al., "Clinical Trial in a Dish Using iPSCs Shows Lovastatin Improves Endothelial Dysfunction and Cellular Cross-Talk in LMNA Cardiomyopathy," *Science Translational Medicine* 12 (2020), doi: 10.1126 /scitranslmed.aax9276.

23. J. Corral-Acero, F. Margara, M. Marciniak, C. Rodero, F. Loncaric, Y. Feng, A. Gilbert, J. F. Fernandes, H. A. Bukhari, and A. Wajdan et al., "The 'Digital Twin' to Enable the Vision of Precision Cardiology," *European Heart Journal* 41 (2020): 4556–4564.

24. M. Abulaiti, Y. Yalikun, K. Murata, A. Sato, M. M. Sami, Y. Sasaki, Y. Fujiwara, K. Minatoya, Y. Shiba, and Y. Tanaka et al., "Establishment of a Heart-on-a-Chip Microdevice Based on Human iPS Cells for the Evaluation of Human Heart Tissue Function," *Scientific Reports* 10 (2020): 19201.

25. H. Dunning, "Earliest Heart and Blood Discovered," April 7, 2014, Natural History Museum, London, https://www.nhm.ac.uk /discover/news/2014/april/earliest-heart-blood-discovered.html.

INDEX

Page numbers in italics refer to figures.

.